Praise for *Our Last Best Act: Planning for the End of Our Lives to Protect the People and Places We Love*

"A living testament to a conscious life because it does not shun death, but embraces it. Every household should have this essential book in their library as a reference point and a point of revelation, pragmatic and visionary at once. Freedom lives within these candid and composed pages. *Our Last Best Act* is a gentle and compassionate bow to the Earth. Let our decomposition be our resurrection."

—Terry Tempest Williams, author of *Erosion: Essays of Undoing*

"*Our Last Best Act* will change your death, and maybe even your life."

—Bill McKibben, author of *The End of Nature*

"Mallory McDuff delves into the difficult topic of death with grace and aplomb, showing how dying, and the choices we make in the aftermath of death, is a mutual and intimate experience that extends across generations."

—Devi Lockwood, author of *1,001 Voices on Climate Change: Everyday Stories of Flood, Fire, Drought, and Displacement from around the World*

"A worthwhile and decidedly pleasant book that aids a valuable purpose in our complicated times and truly speaks to one of the deepest responsibilities of being human: caring for and burying our dead."

—Elizabeth Fournier, author of *The Green Burial Guidebook: Everything You Need to Plan an Affordable, Environmentally Friendly Burial*

"Mallory McDuff's commitment to explore in depth the options available to us when we die, and to complete her own end of life plans, in alignment with the earth, and the realities of climate change, is admirable in and of itself. *Our Last Best Act* exemplifies beautifully how one individual's intentions and courage can model new ways of being for others and help transform society for the greater good."

—Lucinda Herring, green funeral director, home funeral guide, and author of *Reimagining Death: Stories and Practical Wisdom for Home Funerals and Green Burials*

"This book is both practical and political, earthy and spiritual. It will help you think about not only your death. but also your life, with fearlessness and clarity."

—Lauren F. Winner, author of *Wearing God*

"More than a fascinating manual on how to plan for death, *Our Last Best Act* is a loving invitation to courageously face a great and sacred responsibility."

—Janisse Ray, author of *Wild Spectacle: Finding Wholeness in a World Beyond Humans*

"Mallory McDuff's heartfelt saga of discovery will leave you inspired by death and less afraid of it."

—Amy Cunningham, funeral director and owner, Fitting Tribute Funeral Service

"Mallory McDuff tells the story of the journey we are all on with wit, intention, insight, and tantalizing curiosity. Her vibrant description of conservation burial captures the essence of our mission that includes family harmony, personal resonance, and environmental justice when contemplating one of our most potentially impactful life—and death—decisions."

—Lee Webster, former President of the Green Burial Council International and the National Home Funeral Alliance, and co-founder of the Conservation Burial Alliance and the National End-of-Life Doula Alliance

"It has been a blessing to read *Our Last Best Act* while my attention has been on my own mother's death. I urgently want friends and family to read it as they plan for their peaceable end as a part of healing this earth."

—Brian Sellers-Petersen, Agrarian Missioner for the Episcopal Diocese of Olympia and coordinator of Good News Gardens for The Episcopal Church.

OUR LAST BEST ACT

PLANNING FOR THE END OF OUR LIVES TO PROTECT THE PEOPLE AND PLACES WE LOVE

MALLORY McDUFF

FOREWORD BY BECCA STEVENS

BROADLEAF BOOKS
MINNEAPOLIS

OUR LAST BEST ACT
Planning for the End of Our Lives to Protect the People and Places We Love

Cover design: James Kegley
Cover image: MUSTAFFA KAMAL IKLIL/iStock

Print ISBN: 978-1-5064-6446-6
eBook ISBN: 978-1-5064-6447-3

Author's Note: This book has been written with the help of interviews, secondary research, field visits, social-media posts, and observation. In some cases, names and identifying details have been changed to respect and protect the privacy of others. Dialogue has been reconstructed based on extensive notes taken during interviews and my own recollections. In the text, I used the word "earth" to encompass multiple meanings (soil, ground, planet, home) and chose not capitalize it. I have tried my best to tell my story while honoring each of the many stories featured in these pages.

For Lyn and Margaret,

Friends forever

CONTENTS

FOREWORD

About ten years ago, I was walking with a friend in the woods outside Nashville, Tennessee when we passed indentations in the earth around some paw paw trees. My friend pointed toward the slight depressions in the ground: "Those are where former slaves were buried before this became a natural area," she said.

I'm a pastor who has worked for three decades with women who have survived the streets, jails, and trafficking, yet these words made me feel weak in the knees. Those unmarked sunken spaces in the earth seemed sacred, and I wanted to learn more.

Soon I became part of a small group that founded the first conservation cemetery in Tennessee, where natural burials help to protect the land and honor the dead. Now Larkspur Conservation spans more than four hundred acres of green hills and valleys, and we have laid many to rest in a way that honors the earth, our bodies, and the sweet burial rites from long ago. It has been a gift to learn that even in death, we can find healing and authenticity in ritual and burial practices.

As a young priest in 1996, I began serving women survivors of trafficking, trauma, and addiction. From my own history of childhood trauma, I knew that for healing to occur, caring for bodies was integral. I founded Thistle Farms because I believed with all my heart that we can't love one another without caring for the wellbeing of our bodies, our economies, and our spiritual journeys. The community offers this wellbeing to individual women and challenges cultural myths about why women are on the streets and what it takes to welcome women home. From the beginning, we believed that love heals. Women survivors first manufactured and sold healing oils because they were profitable, theologically grounding, and nurturing for both the producers and consumers. Now the healing oils and other body products sold across the country have helped thousands of women find sanctuary.

All justice work is connected. If it is not healing to our bodies, it is not healing to our spirits. If it is not healing to the earth, it is not good for us. This is true in our lives as well as our deaths and reflects the central themes of *Our Last Best Act*. With her daughters, Mallory McDuff spent a year researching options to revise her final wishes in a way that valued the climate and community, as well as her parents whose sudden deaths changed her forever. Through first-hand research and intimate storytelling, she explores how we can build those connections between life, death, and earth in our own communities, especially in a climate crisis.

As a mother and teacher, Mallory has done the heavy lifting to help us make informed choices for the end of our lives. Her story reminds me of the first year the conservation cemetery was open when we laid to rest a woman who had been a resident at Thistle Farms. She had known suffering, prison, and alleys for much of her life, but we sprinkled dirt on her grave on a hillside overlooking wildflowers, where two deer appeared before leaping back into the woods. Like Mallory's journey, this too was a last best act on sacred ground with love at the center of it all.

Peace and love,
Becca Stevens

Chapter 1

MATTERS OF LIFE, DEATH, AND EARTH

Finding a Map for the End of Our Lives

"I HOPE TO BE around for a long time," my father said, "but I've written my funeral plan so we're all prepared."

That morning, he'd cycled to the Bee Natural Farm in our hometown of Fairhope, Alabama, where he volunteered in exchange for organic vegetables. My dad was sixty-two but could pedal faster than his four middle-aged kids. He had gathered us at our childhood home to share his goal of having a burial that relied on family and friends, not a funeral home.

After almost four decades of marriage, he was learning to live alone. My father, a retired IBM salesman, wanted to make sure he—and we—were ready for his death when the time came. From my seat in my mother's comfy reading chair, I could see two single-spaced pages of typed instructions in his hands.

"First, I'd like my body to rest in the bed under Mom's quilt," he said. "Then you can wrap me in linen tablecloths and place me inside the casket. I've talked to my friend Jeff, who'll build my pine casket if I can't do it myself."

With smile lines etched on his cheeks, Dad reminded us that embalming—what he called "filling a dead body with chemicals"—wasn't required by any state law. Nearly a decade ago, my parents had purchased two plots at the nearby cemetery, adjacent to the post office. After we were grown, they trained for their long-distance hikes by putting bricks in their backpacks and walking through suburban woods to the burial ground with its expansive oaks.

My dad had studied his cemetery contract. Like many small-town resting sites, this one didn't require a vault, a 2,500-pound box of reinforced concrete that lines a grave to keep the ground level and maintain ease of landscaping. Most larger lawn cemeteries require a vault, which can cost thousands of dollars.

He wanted his body to stay at home until the time came for a twenty-four-hour vigil at St. Paul's Episcopal Church, followed the next day by the burial. Dad requested plenty of shovels nearby, so old and young could fill the grave with soil. And he'd written a playlist for his bluegrass gospel band, starting slowly with "Amazing Grace" and ending with the upbeat rhythms of "I'll Fly Away." He hoped his band wouldn't be around when he died, as the bass player and vocalist were several years older than him.

"Use some of the money we've saved on the funeral to hire some good local musicians," he said.

The level of detail felt suffocating.

At thirty-eight, I wasn't ready to plan for his death, not when I needed him as a grandparent to my children, as a parent to me. A month before our conversation, Dad had lost his cycling partner and soul mate when my mom, who was only fifty-eight, was hit by a teenage driver. They'd cycled together to an early yoga class to practice sun salutations and savasanas and then biked to the farm, where my father planned to work for the rest of the morning.

My mom took a detour to retrieve the warm gloves she'd forgotten at yoga, but she never returned to the farm to pick up the fresh produce for her bridge club. At their house, the dining room table was already set with her white linens, sterling silver, and cloth napkins.

My mother was killed, her neck broken by the harsh collision of her body with a vehicle driven by a young man. Sometimes I imagined the impact—hard metal on soft skin, a bike and the body of my mother, my emotional cadence, thrown to the ground.

I was still adjusting to my father's daily phone calls and the crumbs on his kitchen counter, which my mom would have wiped clean in a heartbeat. After supper when I visited, my dad would invite me to play cards, a proxy in my parents' nightly game of rummy to decide who had dish duty. He counted cards. She didn't. And of course, my mother usually won.

My father anticipated a sustainable death as a part of a vibrant life. When I was in middle school, Dad built a prototype of his own casket, and my mom kept her jewelry in the pine box the size of her palm. He wore a suit and tie to sell computers to hospitals and universities but chopped wood on the weekends to heat our house. My parents hiked the Appalachian Trail, Pacific Crest Trail, and most of the Continental Divide Trail, aiming to leave no trace in the wilderness, and he aspired to the same ideal in his death.

Like an Eagle Scout collecting badges, he found as much joy in planning for a goal as in achieving it. In anticipation of hiking the 2,100 miles of the Appalachian Trail, he and my mom spent months preparing ready-to-cook, one-pot meals in Ziploc bags and packing more than 700 breakfast and candy bars for snacks along the way. The living room served as a staging area for the boxes mailed to stops on the trail.

In a column for the local newspaper, he wrote about the mixed reactions to their upcoming hike: "Our children seem proud of us, and our preacher says that's something he would like to do," he said. "Our parents, on the other hand, wonder where they went wrong, and the bridge club wonders why anyone would want to do something like that."

He concluded that they were moving forward toward their dreams without always understanding the reasons why. I wonder if that tension between holding the present and the future had something to do with their sense of the ephemeral nature of being. For them, joy had nothing to do with ease, as poet Ross Gay has said, "and everything to do with the fact that we're all going to die."

As I later taught environmental education to college students, my dad's search for a death that used minimal resources showed me how our common fate of death—and planning for it—could be an integral part of all our lives.

In small-town Alabama, my parents raised four children in a 1970s suburban house at the end of a cul-de-sac. They raised us with a respect for the outdoors, faith, and cost cutting. We kept boxes of freeze-dried food in the basement in case of a disaster, and the six of us biked and camped together as a family. During the forty days of Lent and before the advent of recycling, our family gave up trash—aiming for a waste-free household as a spiritual discipline. With an enthusiasm that could bleed into self-righteousness, my father was hell-bent on conservation of resources and money. My mother was an ally who modified his visions to make the plans more practical. She was our glue and our grounding.

Two years after my mother died, I picked up the phone when my sister called with news I couldn't fathom. My father, unbelievably, had been hit by a teen driver while cycling to the farm. He'd been wearing a bright reflective vest and riding on the shoulder of a wide street, safety precautions he'd adopted after my mother's death. But his neck was broken, and his life ended. After this hit-and-run incident, caused by a driver under the influence of drugs, my father's body was taken to the coroner's office and then to the funeral home. While he wasn't at home in his bed as he'd wished, the funeral director agreed we could prepare his body for burial.

The time had come to put his plan in place.

Taking deep breaths in the foyer of the mortuary, my sister and I entered the refrigerated room, where he lay on a metal gurney, covered by a plastic sheet. I ran my fingers along the same pattern of wrinkles on his face that would later mark my cheeks. He had only a few scratches on his body. We sang the gospel lullaby "All God's Children Got Shoes" and began to wrap him in the linens ironed by Mom's hands.

With the solid weight of his limbs against my chest, I lifted this compact, lithe man as my sister held his other side. We carried his whole life in our hands and placed his body into the wooden box, built the night before.

Our friend Jeff transported the casket in the back of his pickup truck to the church and then the grave. He'd used old sailing lines from the basement to fashion the straps for pallbearers to lower the pine box into the earth. At the cemetery, my oldest daughter stood by the gravesite with a shovel taller than her head while my dad's band played "Will the Circle Be Unbroken." The grave closed as we all sang his favorite tune, "I'll Fly Away."

His plan had given us traction to move together through his death, a map of what to do next. But when I looked at my six-year-old daughter Maya, shoveling dirt into a hole in the ground, my body trembled, and my face turned red with tears in the summer sun. Four months pregnant with a second baby, who would be named after my mother, I sweated grief and fear. The faces of the people singing around the grave seemed hazy, like identities erased from a video. I didn't know how to be present for my children in the absence of my parents. How could I prepare my daughters to hold both life and death at the same time?

Fifteen years later, I looked out at a parish hall of my Episcopal church, the Cathedral of All Souls, filled with one hundred people in my home of Asheville, North Carolina. I wasn't at a funeral. Instead, I was one of several speakers at a conference to help others anticipate their deaths.

During a workshop about preparing end-of-life directives, a friend who was a social worker, turned to me and whispered, "I need to revise my will, since my ex is still in it!" I hadn't looked at my final wishes since I had drafted the documents almost a decade ago, several years after my own divorce and my parents' passing.

We then heard a presentation on home funerals, keeping a vigil for the deceased at home, rather than taking the body directly to a mortuary. Caroline Yongue, director of the Center for End of Life Transitions, showed slides of individuals who had died, covered with favorite quilts and fresh flowers with family members by their sides. She helped people care for the body using dry ice to avoid the environmental damage of embalming, a Civil War practice that persists today although it deposits toxic formaldehyde into the earth.

"No state law requires embalming," she reminded us, "And it's completely legal to transport a body in your car and even to bury your loved ones on private land in our county."

Her photos depicted heartbreaking but intimate images of a young teenage girl in a white dress helping to carry the shrouded body of her mother to the gravesite. I turned to the friend next to me. Our oldest daughters, now college students, had gone to preschool together. Her eyes, much like mine, were filled with tears.

In my session, I described the steps we'd taken to bury my father without embalming or a vault when we had only three days to plan. My goal was to share my parents' story and describe local options for green burials that restore, rather than degrade, the earth.

"My sister and I thought we were the only graduates of our high school who'd prepared our father's body for burial in the refrigerated room of the local funeral home," I joked.

Someone asked me about the impact of my parents' deaths on my children. My daughters seemed more comfortable watching emergency-room resuscitations on *Grey's Anatomy* than talking about our own mortality, I answered, but they would listen to the stories of my parents. Although my father's expositions on death had felt like background noise to me as a teenager, he set a tone that has influenced my actions to this day. I wanted to do the same for my own children.

At the conference, I also shared what I'd learned about conservation burial grounds like Prairie Creek Conservation Cemetery in Florida and Ramsey Creek Preserve in South Carolina, one of the first modern-day cemeteries working to protect land while connecting families to death and dying. Near Atlanta, Georgia, I had visited Honey Creek Woodlands, a 78-acre conservation cemetery and part of 2,200 acres of the Trappist monastery grounds. A monk took me on a butterfly expedition using field guides in the cab of his pickup truck, and I told him my parents would have loved this melding of life, death, and earth.

I talked about discovering how natural burial can become one tool to conserve land in perpetuity through easements, which are legal agreements to protect land from development. With conservation cemeteries, funds paid to purchase plots went back to ecological restoration of the land, nourished by the decomposition of bodies. This was a circle-of-life revelation to me. Having time and space to anticipate the end of life is certainly a privilege, which I wanted to acknowledge as well. Yet it wasn't until years after my parents' deaths that I thought to align my values and my work with plans for my death.

My father wanted what could now be termed a green burial. As the nonprofit setting those standards in the United States, the Green Burial Council defines cemeteries as green if they care for the dead with minimal environmental impact—and also help to conserve natural resources, reduce carbon emissions, protect worker health, and restore habitat. These burial grounds must also use biodegradable burial containers and avoid toxic embalming. There are several categories of such sites, from conventional cemeteries that allow green burial in selected plots to conservation cemeteries that partner with local environmental groups through easements. As another option, my dad achieved his wishes at a small-town cemetery, knowing he could be buried there without a vault or embalming.

In preparation for this gathering at my church, I'd also visited Carolina Memorial Sanctuary, a relatively new conservation burial ground outside of Asheville that protects and restores the land. Walking in this cemetery felt like strolling in a wooded preserve, rather than a manicured lawn. I had to look hard to even see the gravesites surrounded by native grasses, trees, and shrubs.

In my forties, I'd made a will and directives, but the final wishes I'd written then didn't seem to fit in my fifties, especially given the climate emergency we face. It hadn't even occurred to me that I'd need to revisit these documents over time. I'd thought final directives were like braces, something you did only once if you were lucky. Yet as a teacher and mother, I was surrounded by the next generation every hour of my days, and I wanted a sustainable end of life to help preserve the land around me for their future.

That evening after the conference, I opened the file cabinet in my bedroom and took out the folder containing my will, cremation directive, and advance care directive, which I hadn't touched in more than a decade. It was like trying on a pair of jeans from high school. My sister was listed as the health care power of attorney, although she'd moved from nearby Atlanta to Seattle. I'd completely forgotten my instructions for a party after my funeral with beer and barbeque from Okie Dokie's Smokehouse, a restaurant I loved when my young children needed a quick fix of mac and cheese and ribs on a school night. (It claimed to offer "swine dining.")

Since that time, my two daughters had grown from toddlers to teens, and my hair had turned a soft gray like my mother's. Ten years earlier, I'd opted for cremation for its cost and convenience, but my research had taught me about the fossil fuels needed to burn a body for several hours into the neat bag of ashes and pulverized bones. The ease of cremation still appealed to me, but I'd learned about more sustainable choices to leave the earth in the same way I'd tried, however imperfectly, to live on. As I held the documents, I saw the possibility of planning for death as my last best act for my children in a warming world threatened by more severe disasters, from blazing wildfires to destructive hurricanes.

Sitting on my bedroom floor beside the file cabinet, I remembered the instructions given by Caroline, the director of the Center for End of Life Transitions. She recommended placing these directives in the freezer, where they would be protected and easy to locate.

"It sounds unusual," Caroline told the crowd, "but the freezer is actually a safe place to store your documents. They won't get touched in a fire, and your family can easily find them."

I'd picked up a refrigerator magnet and a Ziploc bag for the documents labeled "Matters of Life and Death Inside." I hoped to have years ahead to prepare my children for my absence. But after revisiting my wishes, I wanted to feel more prepared for the unknowns ahead.

I told my twenty-one-year-old daughter Maya about the option of keeping my body at the house until the funeral.

"Ew, Mom," she responded, "No thanks."

She explained, "If your body was in the house, I would feel like it was haunted. I couldn't sleep."

My fourteen-year-old Annie Sky had a more direct response: "I will sleep in a Motel 6 if that happens. I can pay for it myself."

They couldn't imagine spending a day in the presence of my dead body, and I didn't blame them. Clearly, my journey to revise my own wishes as a single mother would need to involve my children. During our small but meaningful lives, we were in this together, for now and evermore.

My daughters were one of the reasons I wanted to plan for death that might nourish the earth rather than contribute to its degradation. Conventional burials in the United States generate enormous environmental costs, especially from the release of greenhouse gases produced by the manufacturing and transport of expensive caskets, chemicals for embalming, and concrete vaults buried underground. These products aren't legally required but have become standard practice following the establishment of the funeral industry after the US Civil War.

But the tide is turning. Nearly 54 percent of people in this country are considering a green burial, according to a survey by the National Funeral Directors Association. In response, some funeral homes are diversifying their services, such as collaborating with green burial grounds.

Each year in the United States, conventional burials require the production and transport of 104,000 tons of steel, 2,700 tons of copper and bronze, 1.6 million tons of concrete, and 1.6 million gallons of the known-carcinogen formaldehyde in embalming fluids, and 20 million board feet of hardwood. The embalming fluids in buried bodies leach into the groundwater, while pesticides used to maintain manicured cemetery lawns can harm wildlife and water quality as well. The bottom line from these statistics? These burial grounds often function more like landfills than bucolic resting places.

Today more than 50 percent of the US population chooses flame cremation for its lower cost and convenience, just as I'd done in my final wishes. While less of an impact than conventional burial, cremation still requires fossil fuels to produce high temperatures of at least 1,400 degrees Fahrenheit for several hours and releases greenhouse gases and pollutants like mercury and dioxin. As a frugal neat freak, I appreciated the cost and tidiness of cremation for the consumer—an entire body contained in one plastic bag, as easy as picking up a prescription. But my children are coming of age in a world threatened by a climate crisis.

My daughters have watched daffodils bloom in December in North Carolina due to warming trends in these mountains.

In one week, my brother's house in Alabama was threatened by a hurricane while my sister's family was trapped inside due to wildfires in Washington State. Like many, my kids are unsure what they can do on a daily basis to combat extreme weather events and the disproportionate impacts on the poor and communities of color around the globe. Sometimes it's easier to focus on the present: do homework, volunteer, maintain streaks on Snapchat.

But they care. And so do the hundreds of students I've taught over the past two decades. One of my former students, Kelsey Juliana, led twenty other youth plaintiffs in a lawsuit asserting the federal government violated their constitutional rights by encouraging greenhouse gas emissions while cognizant of the dangers of climate change. Young people I know are advocating for carbon taxes, starting organic farms, installing solar panels, and demanding racial justice in their communities. They recognize the colonialist, patriarchal power structures that support our fossil fuel economy. And they are building change through both individual and collective action.

Before the coronavirus pandemic, an average of more than fifty million people died each year worldwide, including three million in the United States alone. The US is also the second-largest contributor to carbon dioxide emissions worldwide. How we handle the bodies of those who die can influence the climate on various levels: by reducing the impact of greenhouse gas emissions, protecting land as a conservation strategy, and creating a spiritual connection to the earth. And each of these three factors has impacts on the world our children inherit.

Importantly, reducing environmental costs decreases the financial burden of burials for families. Forty percent of households in the United States earn less than $50,000 a year, yet the average cost of a conventional funeral in this country is $10,000. As a single parent and a teacher, I saw the price tag of my disposition, or what happens to the body after death, as an economic justice issue as well.

"Our best last act may, in fact, be the simple act of using what remains of our physical existence to fertilize depleted soil, push up a tree, preserve a bit of wildland from development, and in the process, perpetuate the natural cycle of life that turns to support those we leave behind," writes Mark Harris, author of *Grave Matters*.

In light of the political realities that stymied our response to the climate crisis for too long, I often felt overwhelmed. My students and I talked about feelings of "climate anxiety" and "climate grief." In this context, many people want to have an environmentally conscious end to their life, but they don't know where to start. Decisions about what to do with our bodies after death are deeply personal and depend on our culture, religion, race and ethnicity, finances, and more. But in the end, that legacy has impacts beyond our individual lives.

I began to understand those consequences as I devoured Katy Butler's *The Art of Dying Well*, Atul Gawande's *Being Mortal*, and other books about the medicalization of death, which often prolongs life. I pored over memoirs about death: Cory Taylor's *Dying: A Memoir*, Breeshia Wade's *Grieving while Black*, Kerry Egan's *On Living*. Much as my mother's generation took back their right to birth babies without general anesthesia, we can learn to make decisions about our bodies after death—*before* we face the end of life.

This book chronicles my one-year journey to explore choices about my death that could help protect the people and places I love. In that quest, I also sought answers to several questions: Is this the place for my body? Are these the people whose end-of-life work I want to support? Is this a price I can afford? Any decision I made had to promote healing of the land in a time of climate crisis. That wasn't negotiable.

At the end of that year, I aimed to revise my final wishes—written a decade earlier—and include my daughters in those decisions. The timing seemed especially relevant as I approached the age of my mother at her sudden death. For this project, I contained my search to my home region of Western North Carolina, but my findings relate to national and even global trends. The options I explored reflect my background and values, and this book tells a story through that lens. It isn't an overview of the death-care industry or an analysis of cross-cultural funeral traditions. Those books and stories have been written. But my hands-on experiences—and the questions I tried to answer along the way—provide practical lessons for anyone considering end-of-life decisions with impacts on both climate and community.

During one year, I visited funeral homes and crematories, attended home funerals, volunteered at a conservation burial ground, explored neighborhood cemeteries, shopped for coffins and shrouds, and even observed a university lab that handles body donation. These field trips, as I saw them, revealed both costs and benefits of options as I created a "death plan."

Since I'm a horrible multitasker, I immersed myself in learning about death for the year, an obsession that sometimes concerned my youngest daughter. One summer evening, Annie Sky and I attended a death and dying potluck at Holding Space, a nonprofit created by two friends, Erik and Gabriel, who reimagined end-of-life care that would allow people to die in community regardless of their resources. I'd prepared my daughter for our visit with Yvette, a woman bound to a wheelchair and dying from cancer. As we approached the house, she whispered into my ear, "Puh-leeze, don't ask the dying woman about plans for her body. Just leave it this one time." Yvette would die later that year, and my friends would dig her grave and bury her body at the conservation cemetery.

At the beginning of my journey, I soon discovered I wasn't alone. Many of my friends either hadn't documented their current wishes or had shifted their thoughts about dying. One evening, my neighbors Lyn and Morning talked about death over a beer in my rental duplex on campus.

"When I wrote my will, I wanted a sky burial in Tibet," Morning said, describing the ritual of leaving a body on the rocks for vultures to consume. "I loved the idea of being close to the clouds. But that plan doesn't seem practical, so my second option is cremation, like my mom's end of life."

In 2006, Morning and I had taken a group of students to the Everglades for an Outward Bound course. After our canoe expedition, she scattered her mom's ashes in the Atlantic Ocean.

"It was a blustery day," she recalled, "I threw a handful of ashes into the water, and the wind knocked them back in my face." She described a group of men laughing as the ashes detoured into her hair and skin, adding that she chuckled with them. As she spoke, I glimpsed hundreds of fireflies lighting up the night outside my window.

After a pause, I shared my outdated directives: "I included a party with Okie Dokie's in my wishes," I said. "But I'm seriously rethinking cremation, as well as the fried pickles."

"Your cremation directive was a time capsule of your life ten years ago," Lyn added. "It makes sense the vision might change."

We waited to hear her plans: "I wanted to be cremated, like my Daddy, but my plans could change based on what you find this year," she said. Lyn had taken care of her father, a culturally rich and humble Irishman, who lived with dementia in his final years. We'd attended his wake around her kitchen table with plenty of stories and Jameson's whiskey to go around.

Another longtime friend, the Rt. Rev. Brian Cole, the Episcopal bishop of East Tennessee, helped me consider my journey in broader terms.

"How should we die in order to heal ourselves and the planet?" Brian asked. "How are we going to live beyond our death? We're talking about resurrection in the time of climate crisis. We know the planet will probably survive, but we may not. The legacy we leave with our deaths will have impacts for generations to come."

I thought of my father's wishes. What would he say to me now? This journey became an expression of my parents' love. It was also a prayer that our children's lives could continue on this earth when rising global temperatures—and pandemics—and racial injustice—threatened a safe existence. The intersections were everything.

"When Wendell Berry's father died, his funeral was called his last best act," Brian said. "Our deaths could be the most significant gesture we make for the planet beyond ourselves."

Brian shared one meaningful way to think about the end of our lives and the earth: "A new heaven and earth could be a redeemed planet, where God cares for both humanity and creation," he said. "Our last best act could be dying in a way that contributes to that new heaven and earth, not a landfill on earth in a climate crisis."

That's what my father believed. In my conversations with Brian, I came to see resurrection—coming back to life or back into practice—in a way I'd never considered. With each story I heard from others, my parents' love settled deeper inside my body. And writing about their deaths revealed even wider connections to their story.

"I hiked with Annie and the Salesman on the Appalachian Trail," said one man in an e-mail, referencing my parents' nicknames on the trail. "I'll never forget their smiles that warmed us all." Another epic thru-hiker named Nimblewill recalled his shock when he encountered my brother while hiking years after my parents passed. "After regaining the least bit of composure (the surprise moment 'knocked my props' so-to-speak), I thoroughly enjoyed sharing fond memories of past times spent with Annie and the Salesman," he wrote me. "Their gentle kindness set an example for me, which to this day I try to live up to. Your parents' heart was years ago divinely placed in trust, in a chest—yours."

I'd much rather have my mother and father beside me now in real time, but they provided a navigation system for me and our family—and for others. They showed me earth in its many forms: planet, ground, soil, and lastly, tomb.

My dad told me to wrap his body in my mom's linen tablecloths and sing hymns by his grave, even though I couldn't anticipate the way he would die. I began to see how we all can be stewards for each other, carrying our past and present into the future we hope to see. In that spirit, I began this year looking for a map for the end of my life to honor the people and places close to my heart.

Chapter 2

THE DOCUMENTS CLASS

The Paperwork of Death and Dying

"Complete your death certificate for next week's class," the teacher said. "Fill out the top portion only. The rest will be finished after you die."

Glancing around the room, I tried to look composed and accept this first homework assignment like the straight-A student I'd always been. Next to me, a woman with bouncy brown ringlets sat on a meditation cushion and furrowed her brow while searching for the photocopied death certificate in her three-ring binder. Practicing my active-listening skills, I nodded to the facilitator and gulped my hot ginger tea, scorching the back of my throat.

From the icebreakers to the potluck snacks, this class resembled any of the environmental education workshops I'd attended over the past two decades. But there the similarities ended. This was a "documents course" about preparing for the end of our lives. Would I get extra credit for filling out my death certificate early? I didn't think so. Could I plan for my final disposition in a roomful of strangers? I was there to find out.

After the introductions, I discerned that no one in the small group of six faced an imminent death, but everyone wanted to prepare for that eventuality. Ranging in age from our thirties to seventies, we gathered in the carpeted meeting room—with a sofa, upholstered chairs, and cushions on the floor—at the Center for End of Life Transitions, which is located in a rural area several miles outside downtown Asheville. The facilitator, a former hospice nurse, led us in meditations designed to open our hearts and minds before we examined the worksheets in a three-ring binder to plan for the inevitability of death.

Preparing for death means completing paperwork, the instructor said, but also cultivating a practice of accepting the unknown: we might not die in a way we anticipated. If we want a home funeral, he said, we might die in a hospital bed, surrounded by buzzing lights that never turn off. As I would later see during the COVID-19 pandemic, we might not even have the option of dying in the company of those we love.

Preparing for the end seemed similar to planning for the beginning. As an expectant mother, I had written a birth plan, so when complications came up, the intentions helped everyone stay on the same page. During this course, I hoped to figure out which documents I needed to revise after researching options for the disposition of my body, a final ending that could help to protect the earth.

Yet the day after my first death class, I got flustered—heart racing and F-bombs dropping—as I juggled the mundane task of finding a ride home for my youngest during a conflicting meeting at work. While I rarely lost my shit on the outside, I carried internal angst like a rolling suitcase by my side. How could I stay open to the changing circumstances of my own death when, in my daily life as a mother and teacher, the dropping of one small shoe could throw me? What if I got dementia, couldn't afford health care, and a green funeral was the least of my children's concerns? ("Just cremate her body anyhow," I could imagine them saying.) What if my daughter Maya's fear came true: that preparing for my own death might actually hasten it?

Regardless of how I would die, the purpose of this course was straightforward: to complete a final set of papers needed to plan end-of-life logistics. As a result of this workshop, I could revise my after-death care directive (wishes for the care of the body after death), as well as my North Carolina advance directive, which addressed care leading up to death. At the end of that year, after weighing my options, I hoped to notarize a new set of end-of-life documents. For the purposes of this book, however, I was most concerned with the sheet of paper entitled "Declaration Regarding Disposition of the Physical Body after Death." Those words would communicate the wishes for my body after I died—wishes I would express once I had explored choices that were sustainable for the climate and my community.

Before the course, I'd reviewed my copies of these documents from ten years ago. While my feelings about cremation had changed, I agreed with most of my other plans, especially about not wanting medical intervention to extend my life. All these choices were interconnected, of course, but the primary decision I wanted to reconsider was what happened to my body after death, which would be my last best act.

I'd first learned about the Center for End of Life Transitions from its founder Caroline Yongue, who told me the courses offered by the center had the potential to change my life. Similar organizations and workshops focusing on death literacy and after-death care exist in communities across the country and world.

Before I began my research, Caroline helped me think about people to interview and places to visit as I revised my final wishes. "Sign me up," I said. "I'm ready to be transformed!" Like my father, I loved a good self-improvement project with accountability and concrete goals.

"I don't know why, but I'm a little anxious about interviewing Caroline," I told my daughter Annie Sky, describing my plan to spend the afternoon with the director of the Center for End of Life Transitions and Carolina Memorial Sanctuary, the nearby conservation cemetery she'd founded. That morning, I'd woken up in a state of diffuse fear, a sense of worry around my capacity to manage the many tasks in the day.

My teenager looked me in the eye: "Maybe you don't need to try so hard," she said. And I knew she was right. Trying to impress a Buddhist monk, death midwife, and former hairdresser who sold her house to purchase land for a conservation cemetery was a true koan, a riddle without an answer. But I'd been a fan of Caroline's for more than a decade: she had established a green burial ground where I might decompose someday. As a thank-you gift to Caroline, I'd wrapped a bar of handcrafted lavender soap I'd been saving in the back of my dresser drawer.

Driving into the rural outskirts of town, I passed a Pawn and Gun Stop before turning left into an oasis of hardwood trees, crossing a threshold from the city onto a windy country road. Within a few minutes, I found the Annattasati Magga Sangha and the Mineral Dust Buddha Hall, a rectangular building that looked like a renovated garage. When I peeked into the window, I saw a statue of a Buddha next to the Five Wishes pamphlet—an end-of-life planning document—inside a glass china cabinet. The meditation hall was surrounded by lush beech trees, oaks, and walnuts, like a temperate forest in the humid summer air.

I knocked on the front door of the adjacent building, where Caroline worked and lived. A sign instructed visitors to keep the cat named Blueberry inside.

"Oh shit," I thought, "A cat." While I'd packed a gift, notepads, and pens, I hadn't considered bringing allergy meds.

She opened the door with Blueberry in her arms and her dog Jasper at her heels.

"I wanted to bring you something beautiful," I said, handing her the gift, my face turning red from overeagerness. Thanking me, Caroline sat down on the couch with the cat in her lap and curled her legs underneath her body. We began to talk.

"My experience," she said, "is that we are all trapped in what information we've received about death in our lifetime."

"We check boxes that we want to be an organ donor, but we don't really know what that means," she continued. "We make decisions about the end of our lives, like if we want hydration or nutrition, but what does it actually mean? By gathering with friends and loved ones and talking about these decisions, maybe we will make informed decisions."

I told her my own final wishes for cremation weren't up-to-date with my current knowledge of options, especially with respect to my impact on the earth.

"My initial documents said that I would be cremated too, unless there was a conservation cemetery located nearby," Caroline admitted. "Now we have the green burial ground, but I just picked out a site at the cemetery last Friday."

Her story confirmed for me that directives can evolve over time.

"What's the most important first step people can take regarding end-of-life care?" I asked.

"The most useful thing is to do your darn documents," she said. "Do the will. Do the documents that affect your body. We know we are going to die. As a responsible body owner, make a plan. If we go out of town for several weeks, we arrange for our pets, our bills, our yard. But we fail to take care of the longest trip we'll ever make. So just do your darn documents."

"It's hard to envision ourselves dying," she continued. "We haven't been dead. We think we have time. But the people who plan are able to live their life more at ease."

The cat purred and rubbed against my legs, while I tried to nudge its furry body off mine with subtle movements of my feet. As we shared our own personal stories about death, I felt my eyes welling up, made worse by my reaction to the animals. Soon I was sneezing and reaching for Kleenex, trying to control a full-body allergic reaction to pets in front of a self-contained, animal-loving Buddhist.

"What were your childhood experiences around death?" I asked, steering the conversation away from myself.

"I didn't have much experience with death when I was young," she said. "My grandfather died when I was thirteen years old, but my parents didn't tell me he was dying. At the funeral, they put me to work in the kitchen, but I was so sad and wanted to be with my mother. Later I felt some removal from death, even when I saw roadkill. I had an aversion to it."

Despite Caroline's lack of experience, her Buddhist teacher observed in her an ability to overcome fear and asked her to help with the deaths of members of the sangha. Caroline started by practicing and assisting as a cat's life ended. She spoke about watching the spirit of the animal leave its body.

"I wasn't sure what to do, but I started reading *The Tibetan Book of Living and Dying*," she said. "I was a hairdresser for thirty years, and then my friend's mother had cancer."

Caroline talked with her about what she wanted to happen when she was dying. "We even put a sign on the door, expressing what she wanted from others—how people should behave in the room," she said. "She didn't want people wailing and crying a lot, so her siblings did that outside the room. They knew her wishes because we asked."

To do this work, she said, you need the capacity to hold space, to step past fear.

"I know when I'm walking into a home where someone is dying that I need to leave Caroline at the door," she said. "My job is to hold space for the family and the person who is dying. Not everyone can do it, even if they are trained as a death doula."

"It all seems overwhelming," I said. "Where do I start? I don't want my death to harm the earth, and I also want my daughters to be connected to the end of my life. But they will need some help."

"Start with your beliefs before you create a plan," she said. "What are your beliefs about death? What are your beliefs about what happens after you die? There are implications for a body that is composted, buried, cremated. You may feel it's important to see the body being cremated, so you would pick a crematory where your family can view the retort."

"What do you believe?" she said again. "That's the place to start."

Like a frenzied student, I took notes in my journal, fearful of missing her words and ideas, as if they held some key to the research. But I didn't want to admit I'd never thought much about what happened to the body and the spirit after death. I wasn't sure I held any beliefs about the soul after death—except the literal understanding of the decomposition of our bodies. I'd watched the soil above my parents' gravesites sink a bit as their bodies returned to the earth.

"You can see the body change after three days, when many believe the spirit leaves," Caroline said. "That's why you'll hear death midwives talking to the dead. Some people even believe that the spirit is still there for forty-nine days."

"Whoa, that's a long transition," I said. "I don't think I'll fall into that camp. But what if I'm not sure what I believe? Do these beliefs dictate my life in some way?"

"These beliefs affect how you plan for your death, but they can also affect your life," she said. "You create that energy. Can you get to that place where you can die well? Can you create that space for dying so you can be present for yourself? The last emotional state we're in may determine our future state.

"This morning, you mentioned feeling anxious when you woke up, but you figured out a way to calm yourself by calling a friend. If you can't be present in this very moment, how can you be present at your death? Can you direct the course of your life through a stressful situation? Can you hold steady as the body is falling apart? With a spiritual practice, you can strengthen your own reaction to the unknown."

In my core, I wanted to be someone who could respond to uncertainties, but I wasn't there yet. Caroline's direct questions felt like an outfit I wasn't ready to wear. During the workweek, I tended to wake up anxious, scroll through social media, and feel guilty about not meditating. Then I'd run or do yoga, teach classes, engage in meaningful conversations with students and friends, cook dinner, and settle down again before bedtime.

During the end-of-life classes I was taking, I began a simple contemplative practice of keeping a gratitude and regret jar, which we'd practiced in the classroom. It had seemed to me like reflection for dummies—those like myself who couldn't commit to a daily thirty-minute practice—but I had to start somewhere. Even after my mother's sudden death, my dad began the hard spiritual discipline of learning to live alone, but he never stopped feeling thankful for her life. It was almost otherworldly to witness his ability to hold love in the center of tragedy.

I knew one thing for sure: if I were a high-strung fifty-something single parent, I might become a basket case of an old lady. I imagined myself in my eighties screaming at my middle-aged daughters to take off their shoes before coming into my tiny house parked in their backyard because I hadn't planned for retirement. My mother had prayed and meditated by a single candle every morning before the family woke up, while my father started his meditation practice in his early fifties, about my age. Now their lives and deaths were leading me down this unfolding path.

After speaking with Caroline, I realized I couldn't have completed this project in the months right after my parents' deaths. I would not have been able to evaluate options for disposition of their bodies—and find some peace in the journey—in the seventy-two hours after they died. At that time, I was in a fog, only guided by trying to follow the written directives. The curiosity I now found in talking to people about funerals, coffins, shrouds, and human composting arose only because I had the privilege of time to plan (or so it seemed before the pandemic, when so many of us were reminded of the possibility of so little time).

My research on death and dying felt like planning for a faraway future—until someone I loved faced the end of their life during the pandemic's first year. When an exuberant artist and colleague named Lara shared her diagnosis of terminal metastatic uterine leiomyosarcoma, I first thought of her two young children, not about how she might sculpt an urn for her ashes or find a site at the Warren Wilson Cemetery. Next, I thought about her art students reeling from this news. When I emailed one of her graduating seniors, he responded within five minutes: "The world seems heavy and filled with illness—contrary to the hope and optimism one expects to feel in their final semester before producing a spring art show. She really has been like a second mom to me, so it seems too cruel, too unreal, hard to process. If someone with infinite positivity & joy can be stricken with cancer, it makes me wonder about the rest of us."

In an update on Facebook, Lara wrote, "My oncologist told me that my job now is to live life as stress-free as possible and his job is to help me make it as long and healthy as I can."

Later, she posted a photo of a bright mural she'd been commissioned to paint months before in the parking lot of the hospital: "We came here today to figure out what's next for our family," she wrote as a post on Instagram.

Seeing Lara create art, teach students, and cherish her children—all on social media—taught me and so many others something about valuing life, which goes beyond the logistics of planning for the end. I couldn't know what happens to the soul after we die, but at that moment, I believed in loving her and her children any way I could, even from a distance in a pandemic. As another friend wrote, "She loves with a heat you can feel from a mile away." Maybe this was what I believed: that love continues long after our bodies are buried, cremated, composted. And if my plan reflected my values for the climate and community, I might have more energy to love my people with a heat they can feel from a mile away, even when I am gone.

In my death course, it was time to complete my first homework assignment (beyond the gratitude and regret jar). From my binder, I pulled out the photocopied form, on which "North Carolina Department of Health and Human Services, N.C. Vital Records, Certificate of Death" was written in bold at the top of the page. Our facilitator had told us that the one piece of information families had difficulty finding after someone died was the county of their birth. (I realized my children didn't even know I'd been born in Savannah, Georgia.) I filled out my name, date of birth, gender, occupation, Social Security number, ethnicity, level of education, and names of my parents. But I had to google the name of Chatham County, Georgia, my birthplace.

After my death, others could add the rest of the information, including the method of disposition (burial; cremation; donation; entombment, meaning above-ground burial, like in a crypt or mausoleum; or removal from state) and place of disposition. There was a box to check the manner of death: natural, accident, homicide, pending, suicide, or cannot be determined. In North Carolina, the cause of death on a death certificate could be determined by a physician, a physician's assistant, or nurse practitioner, which includes hospice medical staff. Often, a funeral director files the death certificate with the county, but the family also can submit the document with the appropriate signatures within three days of death.

While North Carolina uses an electronic death registration system, a family can still submit a paper copy. If the deceased was at home, the state also required filing a notification of death with the registrar's office within twenty-four hours. Filling out the death certificate felt similar to the field trip forms I signed every year for my children in school. But this was a lengthy field trip to an unknown destination.

After placing the paper in my three-ring binder at home, I checked Facebook. A friend had posted about an app called WeCroak: "It sends me a reminder five times a day: You are going to die." She signed up to remember her own mortality, like a Fitbit calculating her steps to the end. "In a weird way, it's almost calming to remember," she wrote.

I copied the link to WeCroak and forwarded it to my friend Lyn: "No," she wrote back. "Stop now. This is too much. Do NOT sign me up for reminders throughout the day of my death. Thank you very much!"

We prepare for the great unknown of death by completing something we often dread in life: paperwork. The facilitators with the Center for End of Life Transitions emphasized the importance of mindfulness as people considered filling out the forms, but at the end of the day, as my father well understood, decisions about our death require documentation. As Caroline had said, "Do your darn documents!"

The list of forms was long: North Carolina advance directive, which includes a health care power of attorney and a living will. (The Five Wishes document is another version detailing the type of care and intervention you want or don't want leading to your death.) The only critical change I needed to make to this paperwork was to update my health care power of attorney to include my oldest daughter, Maya, and my friend Lyn as alternate agents, since my sister had moved across the country.

It was hard to believe how time passed: my now twenty-one-year-old had been in elementary school when I first completed the documents. In addition to the forms, I also typed up an inventory of essential information, from bank accounts to computer passwords (which I usually just stored in my head and recreated every time I forgot them).

The still-missing puzzle piece for this story was the after-death care directive: "Declaration Regarding Disposition of Physical Body after Death." So many words for answering one direct question: Where will my body go after I'm gone? As I placed my completed death certificate into my binder, I couldn't anticipate that months later, the courses facilitated by the Center for End of Life Transitions would end with the onset of the pandemic. In emails, rather than in person, the facilitator would finish sharing the documents with me and the other students. As writer Terry Tempest Williams said, "If the world is torn to pieces, I want to see what kind of story I can find in fragmentation." My piecemeal journey to answer the question was about to begin.

Chapter 3

INNOVATIVE UNDERTAKERS

Lessons Learned from Funeral Homes

GROWING UP IN THE Deep South, I'd driven past hundreds of funeral homes, but I'd never been inside one until after my father died.

"Do we even know where his body is right now?" one of my siblings asked as we sat in the living room of our childhood home.

It was close to midnight the night he'd been killed while biking on the shoulder of the road in Fairhope, Alabama, in the summer of 2005.

"The police took his body to the coroner's office," my sister said. I imagined a middle-aged guy pronouncing a body dead or alive, the way a character on a crime show would.

"He's there because the driver fled the scene," my brother Wilson said.

"What about Dad's final wishes?" my sister asked. "He wanted his body at home, and clearly that's not happening."

We knew this wasn't the death my father had anticipated, much as he hadn't expected to lose our mother in a cycling accident two years before.

I held his two-page directive in my hands.

In spite of my fatigue, I could hear him reading aloud his wishes: a casket made by his friend Jeff if he couldn't build it himself, his body wrapped in linen tablecloths, and his bluegrass band by the gravesite.

"If we did one thing on the list, that might be enough," I said.

"We could call Jeff about the casket," my brother said.

I inhaled more deeply, as if I'd just discovered pockets of oxygen in the room. We had all been caught midbreath by this tragedy of my larger-than-life father gone, so soon after my mother's death. She'd been our emotional anchor. He was the helm of logistics. Both were gone. Our kids, sleeping in our childhood beds, would wake at sunrise.

"We could ask the band to play," my sister said.

"And maybe wrap him in Mom's linens," I said.

He'd wanted his death to nourish the land, give something back, and engage his family and friends. Fifteen years later, I would search for that same sense of purpose to give my daughters.

The next morning, my brother woke early to ask Jeff to build a casket within twenty-four hours.

"Turns out he has the perfect piece of pine," Wilson told us.

"Let's check that item off the list," I said.

We'd learned my father's body had been transferred from the coroner's office to a local funeral home. By this time, my other brother, Laurence, had arrived, and we decided to meet the funeral director to ask if we could prepare the body for burial.

He didn't want a hearse or fake-grass carpet at his gravesite. We only needed to keep the body cool until the overnight vigil at the Episcopal church.

* * *

"We'll do what Larry wanted," the funeral director told my brother and me. After my mother's death, her body had been kept in their refrigerated room for two days, so the director was familiar with my dad's minimalist view of after-death care. In her late fifties, my mom hadn't wanted embalming or a vault, but she wasn't as high principled about self-sufficiency as my father. She toed a more moderate line. For her burial, Dad bought a casket from the funeral home and asked them to transport it to the church with the understanding they would leave before any family members arrived.

"Can we place his body in the casket and handle transportation ourselves?" I asked, my voice shaky.

"That's fine," the funeral director said, after a pause. I'd felt numb since losing my father in such a sudden and violent death, yet in that moment, I sat up taller.

Outside, in the humid heat of the parking lot, Laurence and I high-fived each other as if we'd negotiated a high-stakes corporate deal.

Years later, I would learn we had every right to ask for only the services we needed. We could have taken his body home, but we weren't prepared to orchestrate a home funeral. While we hadn't heard the term *green* or *natural burial*, my dad had taught us what it meant: burying an unembalmed body in a biodegradable container and placing it in the earth without a vault.

That afternoon, I returned to the funeral home with my sister as we carried white tablecloths in our arms. The undertaker opened the door to the cool room, where he left us alone to wrap my dad's body in my mother's linens before placing him in the casket.

I waited a few years after my parents' burials to write my own final wishes for what seemed simplest at the time:

cremation, a service at my church, followed by barbeque and beer. Yet when I approached my mother's age when she was killed, I watched my students grappling with the climate crisis, and I'd begun to learn about earth-friendly options beyond flame cremation—options that didn't exist when I sat in an attorney's office to notarize my directives.

My expedition would start with visiting funeral homes, because they are often the first encounter for grieving families. My dad had wanted to avoid what he saw as the excess expense and environmental impact of the funeral industry, but it was the local funeral director who let us prepare his body for burial. In that one act, he played an essential role to "serve the living while caring for the dead," as Thomas Long and Thomas Lynch explain in their book *The Good Funeral*. These coauthors, a seminary professor and a funeral director, believe the rituals and practices around death reflect the soul of a culture, and the body is a crucial part of grieving.

If practices around death reveal the soul of a people, this land where I now live in Western North Carolina is rich with complex histories. On the campus of the school where I teach, there is an archeological site, where my students explore the history of the Indigenous people of this valley, the Tsalagi/ Cherokee, on whose land I have raised my children. The Cherokee and Cherokee ancestors were buried in the villages, often near their homes.

A few miles from the college, students researched and helped to maintain the South Asheville Cemetery, a Black burial ground that was the final resting place for people who were enslaved in this region. During the Civil War, Black soldiers and civilians had to care for and bury Black bodies. Later, Black funeral directors were on the front lines of the civil rights movement, preparing bodies murdered by violent lynching; facilitating "homegoings," as funerals were known; and transporting civil rights leaders in hearses for their safety.

In another cultural practice, one hundred years ago in rural Appalachian homes, if there was a clock in the home, it was often stopped at the time of death. The family then watched the body for twenty-four hours before burial in a local cemetery or the backyard.

Throughout time, our geography, religion, resources, and culture affect how we live and die. Right now, funeral homes remain an integral part of our culture.

I wanted to see what role mortuaries might play in a world in need of more sustainable choices around both life and death. My children could use an education about options: Is this the right place? Are these the people to help? And is this a price we can afford? My death might not go according to plan, I'd learned, but having a plan for my body consistent with my beliefs—as well as an open mind—could be a lifeline and a connection for those who remained.

Through my father's burial, I saw how one innovative funeral director could prioritize sustainability in response to consumer demand. Yet many people like me weren't aware of their right to conduct a home funeral or refuse expensive options at a mortuary. When my friend Lyn's father passed, she walked into the funeral home in his neighborhood to arrange transport of his body and cremation.

"All I needed was to talk to someone about cremation," she later told me. "But the funeral director kept asking if my daddy played golf, which he didn't. He tried to sell me a casket decorated with golfing paraphernalia!"

Lyn knew she didn't need a fancy casket that would only be incinerated. She agreed to pay for an inexpensive cardboard box required by crematories.

"Golf?! Really?" she said.

As a Southerner, influenced by my mother, I thought about what to wear for dress-up occasions like church or a graduation—and now, visits to funeral homes. Before each appointment, I chose an unwrinkled dress and shoes fancier than my slip-on Converses. My daughter Annie Sky had input before I left the house for my first trip into the contemporary funeral industry. "They will be wearing suits and ties," she said. "You should at least wear a bra!" Her impressions came from the shows *Six Feet Under* and *Dead to Me*, and I heeded her advice.

With help from the funeral directors I visited, I learned that death literacy includes admitting what I don't know, not just knowing what I want for my funeral; both are important. I had to recognize my limited experience with diverse funeral practices in this country. During my four years living in Kenya and the Central African Republic in my twenties, I'd attended the funeral ceremonies of many community members. After returning to the United States, I prayed at the funerals of seven close family members who died in the span of a few years. Yet I didn't see an open casket in this country until I was in my forties, at the funeral of a friend's father.

For this research, I visited a range of mortuaries, from full-service funeral homes that offer packages in the thousands of dollars—from $3,000 to $16,000—to others with simple cremation for $995. In the United States, the average conventional funeral and burial costs $10,000, while the average cost for a funeral and cremation is $6,000. My father would have noted with pride that his DIY funeral cost about $150, the price to keep his body cool. Yet I realized that such a "cheap" ceremony would feel like a disgrace for many families, whereas it would have been a badge of honor for my dad.

The city of Asheville has eight mortuaries, including two funeral homes run by Black business leaders. The surrounding region is home to even more mortuaries. In the early 1900s, Black morticians were banned from the National Funeral Directors Association, so they founded their own professional organization, which continues today. Shifting demographics in this region have prompted some funeral homes to diversify their practices: I talked to one funeral director who'd attended workshops to learn about the funeral traditions of Latinx families from different countries.

Today factors ranging from culture to convenience influence the decisions we make after someone we loved has died. Yet comparison shopping becomes a logistical challenge due to time, regardless of beliefs, traditions, or finances. My search was defined by my own values and resources, but I wanted to stay open to the range of approaches toward after-death care.

The Funeral Consumers Alliance is a consumer advocacy organization that provides information to help people navigate pricing in the funeral industry. The executive director, Joshua Slocum, contrasted the process of car shopping with funeral shopping, two big-ticket expenses for families. We would never walk into a dealership, he said, and buy the first car we saw on the lot, yet people often accept the first package deal after a death. With my salary equivalent to that of a seasoned public-school teacher, finances were just as important to me as conservation of the land around me.

On a steamy July morning in North Carolina, I drove from my house to the first funeral home on my journey. The air felt thick with moist heat; temperatures that month would top historic records in this age of climate crisis. At the entrance, I was greeted by a no-nonsense woman in her early forties in a gray jersey dress and black pumps. Morgan Rice would be the hippest, most down-to-earth funeral director I'd meet that year. At the time, I didn't realize she'd also teach me about humility when making plans for death.

"Cosmic!" she'd written in an email, after realizing we had mutual friends who'd started a community-based hospice. Yet she had a quiet reverence for the family members gathered in the front lobby of the mortuary, her steady professionalism balanced with empathic care.

"I wasn't certain I was going to become a funeral director," she later told me, "but now I feel like this is my calling." She was drawn to this work after her grandfather passed away and she planned a family service at a funeral home with musicians "singing him out" from this life into the next.

She decided to attend mortuary school, as her skills in listening and logistics fit well with supporting end-of-life decisions. Morgan was the only funeral director I met who invited me to attend home funerals that didn't require her services. With her help, I got a glimpse into the $20 billion funeral industry but with an eye toward adaptations that were greener and healthier for all.

With its red-brick building and white columns, Davis Funeral Home sat amid the sprawl of Dunkin' Donuts and Bojangles' Famous Chicken 'n Biscuits. Across the street, a structure that looked like an enclosed three-car garage was actually the "care center," the site for cremating, embalming, and preparing bodies for viewing. To date, its facilities hadn't been bought by the megaliths of funeral care like Service Corporation International (SCI), which owns more than two thousand funeral homes and cemeteries across the country.

After I opened the front door, Morgan pulled me aside. "A couple wants to cremate their cat who was hit by a car," she said. "Pets are part of the family too, so I need to meet with them first."

In a typical year, the funeral home handles five hundred cremations or burials. While Morgan made arrangements, I waited in the chapel with its pews, organ, and video screen for use during memorial services. There I met one of the other funeral directors.

"I'm a practical guy, and I don't want a family to spend more than necessary on a funeral. I have to show them all the options, but then I want them to do what's best," he said. Indeed, these costs add up. The $10,000 average price tag for a funeral and burial in this country can become a crushing burden to families who can least afford it.

In recent years some turn to GoFundMe to support funeral arrangements; I'd seen such calls for contributions in my Facebook feed. A 2015 study showed that 17 percent of adults aged twenty to thirty-nine years old had used the internet to request or donate money for funerals. GoFundMe reports that 13 percent of its campaigns in 2017 were memorials, including funerals, the fastest-growing category of funding.

The US funeral industry is required by a 1984 Federal Trade Commission rule to advertise their prices. If you call or visit a funeral home, they must provide a copy of what's called the General Price List. Through comparison shopping at both Davis Funeral Home and its nearby low-cost affiliate, Valley Cremation and Funeral Care, customers could realize savings of nearly $2,000 for basic cremation at the budget facility. The Funeral Consumers Alliance advocates requiring the posting of prices online, but that's not yet the case.

Years ago, I'd read Jessica Mitford's *The American Way of Death*, a 1963 exposé of the funeral industry. Her book, which documented how some funeral directors took advantage of grieving families to sell services, prompted federal legislation regulating the industry. I knew this wasn't the entire story, but the pricing and packages were sometimes hard for me to decipher.

A 2017 National Public Radio investigation of the funeral care industry found "a confusing, unhelpful system that seems designed to be impenetrable by average consumers who must make costly decisions at a time of grief and financial stress." This study found disparate ranges of prices for similar services at different facilities, as well as advertised packages that could prompt consumers to spend $1,900 more than if they were paying separately for services like transport of the body. The commodification of death influenced my father's wish to avoid mortuaries, yet I couldn't forget how our local funeral director in Alabama was there for us after his death. There are good people in the industry serving families at the most vulnerable time of their lives.

"They are going to bring the cat, and you should get a paw print and a lock of hair," Morgan told her coworker.

After her colleague left, she joined me in a pew to talk about the two questions she believes people need to ask themselves before their death.

The first question, according to Morgan, is "What will your final resting place look like? Do you want people to have a place where they go, like a cemetery or an urn in a closet?" She added, "My mom has found urns filled with ashes at Goodwill! Sometimes people don't know what to do with cremated remains."

She listed the primary choices for disposition or final handling of the body: cremation, alkaline hydrolysis (also called aquamation), conventional burial, green burial without embalming or a vault, burial of an urn, and exposure burial, when the body is exposed to the elements outside. By the end of the year, I would explore every one of these options to evaluate which one was best for me.

"What's the second question?" I asked.

"There will be a moment when someone sees you for the last time," she continued. "Is that a family visitation, your home, a conservation cemetery?"

I hadn't thought about the last time my daughters would see me.

"As a funeral home director, we have to show up," she said. "We can help answer those questions. That's why people pay us thousands of dollars. People don't know what to do, and we are there for them."

Our ignorance and fear of death is a recent cultural phenomenon. Until the mid-1800s, families in this country cared for their dead at home before burial in a nearby cemetery, farm, or the backyard. Women bathed, dressed, and prepared the body for viewing in the home, much as my father had wanted his body to rest in his own bed.

In the United States, the Civil War prompted the birth of the modern funeral industry, as prosperous northern families wanted to ship bodies for burial back home. A man named Thomas Holmes developed embalming techniques, practiced on Southern battlegrounds on a wooden plank with two barrels, reportedly at a cost of $100 each. For most people, the transit of Abraham Lincoln's body across the country was their first exposure to preserving bodies with formaldehyde and other chemicals.

Embalming is not widely practiced worldwide, except in Canada and the United States, where it is required only for long public visitations, transit on an airplane, or entombment in a mausoleum. But embalming and cosmetic restoration have become accepted in this country. There are options now for organic embalming with formaldehyde-free plant extracts. In a period of only 150 years, the practice of a family caring for the body at home has been replaced by the funeral home with one-stop shopping for services. With that shift, the cost has become prohibitive for some, creating a crowdsourcing option for the end of our lives.

After we spoke, Morgan left to meet with a distraught woman who needed immediate help, and I reviewed my copy of the prices. The most expensive package for cremation was the Reflections Cremation Plan, which included transportation of the body, embalming, visitation, a photo collage, crematory fee, an urn allowance, and a cremation casket allowance for a total of $8,890. These prices were similar to the packages of other full-service funeral homes I later visited. At this point, even I realized an expensive casket would be burned. Morgan later explained that embalming was rarely used before cremation, except in cases such as delayed funerals or large-scale viewings.

The cheapest option, direct cremation, was $2,870, which included transfer of the body to the funeral home and the services of the funeral director. The prices for Davis Funeral Home were not available online, unlike those for its lower-cost site, Valley Cremation and Funeral Care, where cremation was advertised at $875, the lowest price I found. At the budget funeral homes, the reduced overhead and services contributed to lower prices.

"Some people don't want to even walk into a funeral home," Morgan later said. "So the lower-cost option is for people who want a simple cremation without other services. We're also developing a green burial display at that facility."

At Davis, the Basic Burial Plan was $9,925. If you wanted to be buried in a cardboard casket, your burial services could be as low as $3,485. After a few hours, I'd learned to avoid packages and investigate lower-cost facilities in the same region, since price was important for me. But another lesson was the disconnect I felt between celebrating someone's life—how they'd lived and loved—with an expensive ritual that could harm the land. Some of the practices, like embalming, didn't resonate with my work in environmental education. At the same time, I felt deeply grounded in Morgan's presence, an asset she brought to her work each day. She had something to teach me, I could tell, and our frank conversations would be the way I would learn.

"She's really upset, and I need to finish talking with her," Morgan said after returning to the chapel. She struck an authentic balance between showing care for her client and taking care of logistics.

Before she left, I asked about burial without vaults, one of the criteria for natural burial. The small-town cemetery where both my parents were buried did not require a vault, the concrete box that lines a grave. Morgan leaned over the pew to make her point: "Without a vault, the earth sinks," she said. "People go ballistic if the grave sinks. Unless the family or the cemetery agrees to maintain it, we have to use them."

She had referred clients interested in green burial to the first conservation cemetery in the country, Ramsey Creek Preserve in South Carolina, which protects the land in perpetuity—and does not allow vaults or embalming. Carolina Memorial Sanctuary, North Carolina's first conservation cemetery, is located about thirty minutes away. While my parents didn't have this option, they were able to have a green burial by reading the fine print of their cemetery contract—and communicating their wishes to us all.

"Some people think we are going to take them for a ride," Morgan said when we talked on the phone the next week. "We have caskets that sell from $200 to $6,600. It's not up to me what you select."

She offered to loan me her copy of the book *History of Funeral Homes*, which she'd read in mortuary school. I could tell Morgan was an educator at heart.

"In the past, when someone died, a woodworker built the coffin," she said. "When caskets got fancier, people didn't want to put the casket directly in the ground. So they used a vault. Then people had the expectation that the cemetery would be flat, so cemeteries started requiring vaults."

As she spoke, I knew my father would have agreed that individuals can influence the industry by asking for exactly what they want, especially if they don't want to pollute the land with their death.

"Things will change when customers start asking for more sustainable choices," she said. "If people ask for a pine box, we know where to get one."

During my tour of other funeral homes, I'd seen half-caskets mounted on the wall with a range of price tags: a stainless-steel coffin with white-velvet interior for $3,980, a burial shroud for $390, a $110 cardboard coffin for cremation, and a rental for $880. Much like textbooks in college, you could rent your casket for a public viewing and get buried in a cheaper cardboard model. I considered the choices: my father's homemade pine casket seemed like a good option to me, and tablecloths might work just fine as a shroud.

As Morgan drove through the mountains, the reception cut in and out, and each time, she called me back to resume our conversation.

"The first time I ever saw a body embalmed, I was against the practice," she said. "But then I realized that by flushing out the blood, you can accomplish things with embalming. If someone shoots themselves in the head, you can fix it so their family can see them. You can view a body that isn't embalmed, but you can't carry them all around town for multiple services."

Modern-day embalming involves draining the blood and replacing it with a formaldehyde-based preservative, identified by OSHA as a potential occupational carcinogen. I recall my father telling me that "injecting chemicals into a dead body" wasn't required by law. During my research, I'd also learned the process begins with an incision at the base of the throat, followed by insertion of a tube in the carotid artery. The liquid formaldehyde replaces the blood, which goes down the drain. The funeral director pushes a piece of metal called a trocar into the stomach to puncture the organs, suck out the waste, and infuse embalming fluids.

When I later described that process to my teenager, she put her finger in her mouth, simulating the gag reflex. Through these less-than-pleasant conversations, I wanted her to understand there were choices safer for the soil and for funeral home staff too. She might not remember these lessons, but at least we were talking about the behind-the-scenes realities of death and dying.

"Before we hang up, I want to tell you about discovering the Warren Wilson Cemetery," I told Morgan, who knew the reputation of my college for environmental conservation.

"It's a sweet little cemetery owned by the Presbyterian church on campus," I said. "But I just found out the contract requires a vault!"

It seemed incongruous to bury a cement block at a school where environmental studies was the most popular major and sustainability was central to the mission.

"If Warren Wilson College doesn't let you bury without a vault, that would be crazy!" she said.

I appreciated her practical insights, and I saw Morgan as a mentor, even though she is more than a decade younger than I. She is a woman leading from within a male-dominated death-care industry, and she believes people should be remembered as they lived—like my father desiring his own good death and me yearning to discover my own.

"Mom, you're obsessed with death," Annie Sky said before I left the house the next day to visit another funeral home. But I was in good company. I'd been talking to my daughter about mortician Caitlin Doughty, whom I knew through her books and YouTube videos. Her writing gave me a sneak peek at the industry when I couldn't observe the embalming room or the crematory in action. (I asked permission, but privacy for the families trumped my curiosity.)

She describes the purpose of the industry to "protect, sanitize, and beautify the body." The protection comes with the sale of sealed caskets and vaults, while embalming reflects an effort at sanitizing the dead. Finally, beautifying involves using makeup, hydrating the skin, and masking odors. She shares a story of preparing a baby's body for burial and shaving the soft hair on the newborn's face before applying cosmetics.

Funeral directors talk about "setting the features," which I learned sometimes involves taking small pieces of plastic with spikes and placing them under the eyelids so they remain closed. (Morgan reminded me this wasn't that different from putting in contact lenses.) To close the mouth, a traditional approach is to tie a scarf under the chin and above the head before rigor mortis sets in, when the body temporarily becomes stiff several hours after death. But morticians may use a needle injector to shoot wires through the gums and keep the mouth shut.

"If a body isn't embalmed, we only have a family viewing," another funeral director told me. "We'll clean and dress the body and set the features. We can't take away the death, but we can make it emotionally palatable for them."

In my notebook, I underlined the phrase *emotionally palatable* multiple times.

After visiting several full-service funeral homes, I knew the prices were out of my league, so I decided to check out the budget facilities. A common refrain I'd heard was that the best way to get a better deal in the death-care business was to visit another funeral home down the street. The priests at my church recommended the low-cost Asheville Mortuary Services for cremation because of its proximity and pricing at $995.

Many times I'd passed the gray-cinder-block building but never paid attention to the small sign with green lettering about a half mile from my church. In the parking lot were four white vans, Dodges and Chryslers, looking like gently used family cars. After entering a simple waiting room, large enough for one small couch and two chairs, I was ushered into a space resembling the sparse counselor's office at my high school, although filled with urns and mementos for the deceased.

Within minutes, the funeral director, Stanley Combs, walked into the room in his dark suit and shook my hand. He works at two locations—Asheville Mortuary, which focuses only on cremations, and Asheville Area Alternative, handling both cremations and burials. They were collaborating with the only conservation burial ground in the state in hopes of growing their business for natural burials. In a town like Asheville, known for local food and outdoor recreation, this seemed like a logical move.

"Years ago, Caroline Yongue was doing workshops on home funerals before she started the conservation cemetery," he said. "The other funeral homes weren't entirely supportive of her work."

I'd noticed the pamphlets for Carolina Memorial Sanctuary in the reception area. "To be honest, it's not profitable for us to promote home funerals," he said. "But if it helps the person say good-bye, then that's our job."

His support of these alternatives seemed to hinge on his long-term partnership with Caroline. Yet Stanley said the idea of a home funeral, like having his mother's body in his home, wasn't his "cup of tea." He lowered his prices for transporting bodies to the conservation cemetery because his employees didn't have to provide labor there. The funeral home would drive the body to the burial ground, where the manager at the sanctuary would sign a release and take responsibility from there. The staff at the conservation cemetery would then bury the body with the family participating in the ritual. Later that year, I would volunteer to help with natural burials as well.

"And frankly, most of the families don't want us there," he said. "They are at the sanctuary because of the community and the land."

He was describing my father, who didn't want the funeral home at my mother's burial or his own. Much like Morgan, Stanley didn't see a conflict between doing his job and empowering the family.

With the conservation cemetery, he had agreed to charge a bargain-basement price of only $665, including refrigeration, dressing and shrouding, transport, and the death certificate. If a family chose cremation, then they could pick up the ashes for $995. As with other cemeteries, an individual still had to buy the plot at Carolina Memorial Sanctuary; when I checked, the prices ranged from the lowest price of $1,100 for burying cremated remains to $3,500 for burial of a body. Because of the sanctuary's status as a conservation cemetery, the money supports ecological efforts such as restoration of wetlands and removal of invasive species at the site.

"That's a great price," I said. "What would you charge if I wanted to be buried at the Warren Wilson Cemetery?"

The cost of a plot at the campus cemetery was only $100, an insider's bargain.

"If we handled the refrigeration, transport, and delivery of your body to the Warren Wilson Cemetery, we would charge the same $665, as well as the $995 for the cost of the funeral director services, for a total of $1,400."

Of course, my children could keep my body cool at home with dry ice and transport me to the cemetery on campus at no cost. In that scenario, my burial might be cheaper than my father's, but I needed to give them alternatives. After meeting with Stanley, I decided his funeral home would be an affordable option if I needed refrigeration—or if I had to be cremated for some reason. But as a mother and a teacher, I yearned for Morgan's wry humor and straightforward presence with my daughters after I died. I imagined she would tell me that I'd get what I paid for.

The following summer, as COVID-19 ravaged the country and funeral directors worked on the front lines of the pandemic, I caught up with Morgan on the phone.

"You know, I'm not real thrilled to expose myself and my family to coronavirus," she said. "But I got into this work to serve people, and my full-service funeral home pays me a living wage. I don't think people searching for the cheapest green burial are considering what funeral homes pay their employees. That's a social-justice issue too, you know?"

As I listened, I realized she was right: I'd considered the cost to myself and the earth, but I hadn't factored in the pay for funeral directors.

"You can be an environmentalist, but you need to be open," she continued, "There isn't a right and a wrong choice. Green burial isn't right. And embalming isn't wrong. These are just different options for different families."

I could feel a hot flush creeping up my neck. She'd named my blind spot, influenced by my vocation and my father. It wasn't lost on me that the word *humility* shares a root with humus, meaning earth. This search focused on my own death, but the most sustainable change might come with the intersections between innovative funeral homes and conservation of the earth. And that would take open minds by all.

In a world with choices, we also may have more options than we imagine when faced with the end of our lives—or the deaths of those we love. Families can use an urn from home or buy a cardboard coffin from Costco and bring it to a mortuary—or bypass the industry altogether with a home funeral. This reality may encourage funeral homes to diversify, adding sustainable choices many consumers desire. With increased demand for cremation, the decreased revenue from casket sales has also influenced the need for alternatives. In the United States, casket shipments from manufacturers had an estimated value of $456 million in 2020, down from $1.2 billion in 2000. With this shift, some funeral homes now sell an experience through multisensory rooms that create an atmosphere enjoyed by the deceased, such as a golf course (again with the golf!) or the beach. Others collaborate with cemeteries that offer natural burial, as I'd learned, in a win-win partnership.

After this leg of my mortuary tour, I saw that consumers like me—and ultimately my children—could be advocates for change that protects both people and places. The funeral director in my Alabama hometown might not have accommodated our wishes if my father hadn't laid the groundwork after my mother's death.

"Did you really pick up your father's body and put it in a casket?" my older daughter Maya once asked. "I'm not sure I could touch you when you're dead."

There were days when the weight of his body seemed like a sensation from another lifetime, almost otherworldly. But when I closed my eyes, I could remember entering the cool of the refrigerated room at the funeral home. There I felt his lean body, wrapped in my mother's white linens, his face etched with wrinkles, like mine. I could even imagine my mother's hands folding the pressed fabric after a family meal.

I understood that I carry my grief in my body, a weight pushing me toward healing choices for the children I love and the land I call home.

Chapter 4

DYING AND ITS AFTERMATH

End-of-Life Doulas and Home Funerals

When I first became a mother in my late twenties, I'd seen hundreds of babies in my lifetime—in parks, grocery stores, church nurseries, and living rooms. On my ninth birthday, I even held my younger sister, a three-day-old newborn, and felt the precious weight of her tiny body in my lap. But decades later, when I prepared my father's body for burial, I was a total novice: I didn't know what to expect, and I wasn't alone in my lack of experience. Many in this country will face their own death—or the death of those they love—without ever having seen a body at the end of a life.

I first saw a dead body when I was a college student, working in the medical genetics department of a local university. That experience was notable because of my fear and curiosity.

"Mallory," said Dr. Lustig, the head of the department at the time, "I have a job for you this afternoon." He paused. "I need you to go to the Pensacola hospital to pick up a baby from the morgue for genetic testing."

Would medical personnel really allow a twenty-year-old to just take a dead baby? I wondered as he gave me the details. But I got my keys and headed out the door for the hour-long drive.

Surely, I received some official paperwork, some documentation, but what I recall was the trembling of my hands on the steering wheel of my orange 1970 Honda Civic after I placed the baby, wrapped in blue plastic, on the hot-to-the-touch vinyl passenger seat. It was July, and temperatures soared into the nineties.

Was I supposed to put a seatbelt around the bundle? What would happen if I got in an accident?

As I drove across the bridge over Mobile Bay, the water sparkling in the summer sun, I began to talk to this child in a hushed voice, wondering aloud about her short life in her mother's womb. What sounds had she heard? Had she been given a name? I had no frame of reference for the physical encounter, but I found myself seeking some connection.

When I returned to the parking lot of the medical genetics department, I lifted the plastic surrounding her face, gently peeling up a piece of tape, glancing at this small human beside me and the gray pallor of her skin. With a deep exhale, I sent love to her parents and gently carried her into the cool of the air-conditioned building.

Years later, as a mother with my own children, I'd shared the story of preparing my father's body for burial. I wanted to be sure to give my daughters the option—if they chose it—to be with my body in my home, even if they weren't enthused about the possibility right now. But I also needed to find people who could assist them, much as birth doulas support women bringing their babies into the world. Could I plan a home funeral when I was healthy and alive? Well, I certainly couldn't plan one when I was dead.

Whether I die in a hospice facility, a hospital, or at home, my daughters have a right to make decisions about my body, rather than have it taken directly to a funeral home. My goal was to figure out who can help them exercise that right when they are facing the end of my life.

With my mother's death, I missed that tangible experience, even though I knew my father held her after she died. For me, the encounter wasn't about closure; it was about connection. Those hours with my father's body gave me concrete sensations—touch, smells, sights—to hold onto and return to for years to come. The moment taught me that death ends a life but not a relationship, especially when it's between a parent and child. For all of us, the absence of the presence is forever.

"I'm going to a home visitation Thursday evening, so why don't you come with me?" wrote funeral director Morgan Rice in an email.

"That sounds great," I replied, uncertain of the proper attire or behavior for a home visitation but eager to learn.

"Aditi Sethi-Brown's mother-in-law will be buried Friday at Carolina Memorial Sanctuary," Morgan wrote. "Family and friends are gathering at her house tonight."

As I quickly learned, the death-care community in this region is close-knit. In addition to her work as a funeral director, Morgan conducted community workshops on death literacy. She was friends with Aditi, a hospice and palliative-trained physician and musician, beloved in many circles. By chance, I'd also been asked to volunteer at the conservation cemetery, Carolina Memorial Sanctuary, that same Friday.

When I checked Aditi's Facebook page, I clicked on a video of her mother-in-law, Paula Costello Brown, lying on a bed surrounded by her six sisters. Her body looked small and frail, as if she were already transitioning across a threshold. Each sister leaned toward her to say they loved her and that she was free to go. In a few minutes of footage, I saw a circle of siblings loving a body close to dying. From reading later posts, I learned Paula had advanced pancreatic cancer and died the day before in her son and daughter-in-law's home.

That summer evening, I met Morgan and her mother at Aditi's home, which was surrounded by a wide-open yard, where children played. On the deck, a table held cups and pitchers of iced tea and water. Feeling like an imposter, I followed Morgan's lead as she took off her shoes and walked inside a room that looked like a music or yoga studio adjacent to the house.

Inside that space, Paula's body lay on a massage table covered in tapestries and fabrics. I could see only her face, since a shroud surrounded her and layers of fresh flowers and lavender had been placed on top.

Wearing a cotton tunic and pants, Aditi greeted me warmly, as if I were another welcome guest in her home. She was clearly expecting a baby soon, and her two other young children ran in and out of the room, in the presence of their grandmother's body.

"I'm volunteering at the burial tomorrow," I said, relieved to have some official role to legitimize my presence.

Aditi had arranged six chairs in a semicircle in front of Paula's body, while other family members visited on cushions in the back of the room. Morgan pulled from her bag a makeup brush and a bottle of cold cream, which she used to moisturize Paula's face. From where I stood, I noticed that Paula's eyes and mouth were slightly open.

"I asked Aditi if she wanted me to set her features," Morgan later told me. "But she said it was fine to leave Paula as she was."

I took a seat in front of Paula's body, softening my own breath, my heartbeat slowing. In a corner of the room, Aditi began tuning her harp, while a friend sang, "Make me an instrument of thy peace" in a clear soprano voice.

As newcomers entered the room, Aditi would rise and hug them, folding herself into their embrace, while her husband, Jay, strummed his guitar. I watched as Morgan's mom approached Paula's body and held onto her hands, clasped together on her chest. I assumed someone had placed dry ice under the body to keep it cool. In this reverent space, I knew I wanted the chance to give my own children time to face the end of life with caring adults well versed in the love and logistics of death.

This was my first experience with a home funeral or home vigil, when a body is cared for after death by friends and family. The logistics often involve tasks such as keeping the body cool, using dry ice, ice packs, or a product called Techni-Ice™, which is available online. (*Another illogical reason to keep Amazon Prime?* I wondered.) In most states, these methods of cooling the body—if it's not buried within twenty-four hours—are all legally acceptable. Other tasks the family can assume include filing a death certificate, washing and dressing the body, transporting the body, and helping with the final disposition, such as burial, cremation, or body donation. These roles are also ones a funeral home can take over, depending on the family's needs and wishes.

The week of Paula's vigil, I began to explore options for my own home funeral. My efforts included talking with Caroline Yongue, the director of the Center for End of Life Transitions, which also offers workshops on home funerals. I shared my concerns about wanting to keep my body at home for at least a day without overwhelming my daughters with details like where to buy dry ice.

"Look," she responded, "anyone who can change a baby's diaper can lead a home funeral."

This analogy put things in perspective for me, although I knew it was more complicated.

"Home funerals are legal in every state," she continued. "There are only nine states that require the involvement of a funeral director in some way, but you can still have one. In North Carolina, you don't need a funeral director involved if you don't want one, and you don't need a permit to transport a body unless you are crossing state lines or the body is with a medical examiner."

"So my friends can transport me in a back of a truck, right?" I asked, thinking about Lyn's 1965 bright-blue Chevy truck, once owned by her father and now coveted by students who worked on the campus farm. I thought it could make the one-mile trip from my house to the college cemetery, if I was buried there.

"That's right," Caroline said. "People often drive their loved ones to the conservation cemetery."

"I'd like my daughters to have a home vigil for me," I told her. "But it feels like I'm imposing something on them, because of my experience with my dad's death." I added, "Why do you think home funerals can help families?"

With her blond-gray hair and flawless skin, Caroline looked at me as if she were coaching a young child.

"You can't convince your daughters," she said, "But they are young now. It seems like you're preparing them by having these conversations. The time with your body will give them space to grieve and say good-bye, rather than treating death like some emergency that needs to be whisked away."

When I returned home after visiting Caroline, I spread resources on my coffee table: information from the National Home Funeral Alliance website; an article entitled "How to

Arrange a Home Funeral," from the Funeral Consumers Alliance; and the book *Reimagining Death: Stories and Practical Wisdom for Home Funerals and Green Burials.*

But facing these documents, I felt my heart rate spike, as if I were filing my taxes without a W-2 or teaching a class on a subject I knew nothing about. I imagined my daughters finding these papers full of instructions and to-do lists and then throwing them in the recycling bin after calling 911 and a funeral home to take me away. If it was daunting for me, I was going to have to make this as straightforward as possible for them.

If my death were expected and I were under the care of hospice, for example, they could notify the doctor, report the death, and initiate filing of the death certificate, which I'd started to complete in my class on preparing your end-of-life documents. If my death were sudden, my friends or children would have to call the police or coroner. For me, it was critical to have a plan for this eventuality too, so my body wouldn't get taken away to the nearest funeral home within hours of death.

Once I figured out my final wishes for disposition, my children could in either case take one simple step: calling the Center for End of Life Transitions, which would send a death midwife to assist them. I'd heard their staff make a distinction between *end-of-life doulas*, who help a person transition to death, and *home funeral guides (sometimes called death midwives)*, who are also trained in caring for the body after death. If needed, my daughters could also call a funeral home to help with transport of my body after keeping me at home for one to three days. (If the present dictates the future, I guessed they'd choose the shorter duration).

If I died in the hospital, the center could advocate so my family could take my body home, with the help of a funeral director. Many religious traditions believe the soul leaves the physical body during these days after death, sometimes described as "time out of time." Tibetan Buddhists describe the time as bardo planes, intermediate states between death and rebirth. What happens to the soul in the case of sudden death, like that of my parents? It was difficult to imagine what I didn't understand. As writer Elizabeth Gilbert once posted on Instagram, "Life is a mystery. And there is a period at the end of that sentence."

Even if I didn't comprehend the mystery, I knew it was important to consider each aspect of the time between death and disposition in order to help my daughters. Through searching online, I learned that the state requires filing a "notification of death" with the local registrar of vital statistics within twenty-four hours of taking custody of the body. My death certificate would need to be signed by the doctor or nurse practitioner who determined the cause of death and filed within five days of death. My children would need to have copies of the death certificate certified before being able to arrange for disposition of the body. While the state uses an electronic death registration system, it was still legal to file paper copies, and I had placed a copy of my death certificate in a folder marked "Death."

"How will my children remember all of this?" I'd asked Caroline. "Now I see why people just turn everything over to a funeral director."

"Your children don't have to handle all these details," she said. "They could arrange help from the center. And you could designate a team of people—other friends and family—to support your daughters."

I thought my daughters might prefer to use a compassionate funeral director, such as Morgan Rice or Stanley Combs, to take over any of the tasks and assist them in the process. Each of these contacts would be included in my death plan.

Certainly, that's what happened to me and my siblings, even though we didn't have my father's body at home. My father set a plan in place on the pages of his directives, including the names of helpers like Jeff to build his casket and Anthony to dig the grave. Others assisted without our prompting; for example, my mother's bridge group brought food to the house to feed the extended family.

But it all started with a plan and a conversation, even when we weren't ready to talk. I knew, of course, that I'd need to revisit my intentions as my children got older. If the dying person was the only one who wanted a home funeral, that could be cause for resentment, rather than a healing process. At each step, I could hear my daughters' voices in my head. We would need to keep talking about death over time, and I'd need to stay open to outcomes too.

"This is an ancient practice," Trish Rux told me. "Women have been doing this work for families, congregations, and communities for years."

In her midsixties with reddish-gray hair, Trish projected a midwestern, grounded, can-do presence. She seemed like the type of person who could pack a wound and also manage a plant nursery, and she'd done both in her lifetime. She believes planning for death can help us live a full life. She should know: Trish worked as a hospice nurse at the Veterans Affairs hospital, where she introduced practices including healing touch, meditation, anointing with oil, and aromatherapy into an institutionalized medical setting.

At the time of my visit, she trained people as end-of-life doulas with the Conscious Dying Institute. She also taught a course called "The Best Three Months," designed to help people plan their after-death care and focus on priorities for their life. Later, when the coronavirus hit, she began offering the course online.

When I was pregnant with my daughters, I'd learned about birth doulas, but Trish was a trainer for end-of-life doulas. The role had existed for centuries, but the modern end-of-life doula movement began when a social worker named Henry Fersko-Weiss created the first end-of-life doula program in the United States at a New York City hospice in 2003. He later founded the International End of Life Doula Association (INELDA), a resource for locating doulas within communities. Since then, the National End-of-Life Doula Alliance (NEDA) was created to establish standards for the profession.

I'd wondered how an end-of-life doula differs from medical staff who work with hospice, and Trish explained the roles as complementary but different.

"A death doula is someone who can help you with the planning around death, guide you through the hospice system, assist with the emotional and spiritual work, help your family through grief and bereavement, and also allow you to document your life," Trish told me.

That sounded like a lot of duties to me. In Western North Carolina, there were more than fifty trained death doulas, Trish said, and some charge by the hour on a sliding scale, while others offer their services as a flat fee. When they are able to, many end-of-life doulas volunteer to help families in need.

"It's our role as doulas to show people how they can die well," she said. "We know it's important to see the body and spend time with the body, so we can understand what we too will face one day.

"One point of dying," she added, "is that we are in this together. Your death informs my death."

As we chatted, I knew that's what I wanted—to give my children some glimpse into their mortality but with healing touch, the smell of lavender, a cool towel on the skin. Given that she spoke about the embodied aspects of death with a spiritual understanding, I wasn't surprised when Trish told me she had founded a meditation group at her home.

"I couldn't do this work with death without having a spiritual practice," she said. "I'm not talking about self-care with a bubble bath. For me, it's about developing practices in community that can ground us each and every day of our life—and into our death."

While the bubble bath sounded good to me, Trish's advice echoed some of what I was reading in *Advice for Future Corpses: A Practical Perspective on Death and Dying*. In the book, Sallie Tisdale writes that growing more comfortable with surprises can help us meet the unexpected and unknowns at death. It almost seemed like the unraveling of the birth process, this shutting down of our life.

Trish shared with me one end-of-life practice: the anointing ritual she'd brought with her to the VA hospice, which families could use before death or at the time of death, with adaptations to make it their own. This optional practice, performed by a family member, hospice nurse, or religious leader, involves pouring a small amount of essential oil into the palm, placing a dot on each part of the body, and sharing a blessing:

> May you receive this anointing and be honored and loved unconditionally. As your body is outwardly anointed with oil, so may you be filled with love. May you be held in a state of grace, affirmed in all your goodness, and may you have eternal peace of mind, body, and spirit. We anoint this body that has journeyed through this life on earth.

We anoint these eyes that have seen so much.

We anoint this mouth that has spoken love and truth.

We anoint these shoulders that have borne so many burdens with grace.

We anoint these hands that have worked so hard.

We anoint this heart that has loved so well.

We anoint these feet that have traveled so far. May you rest in peace.

My visit with Trish caused me to reflect on my life, not just the end of it. My mother had opened each day with silent prayer and reading, yet my own heart still tended to race through the motions of my days. I'd started wondering how learning about death might help to quiet me now. I was so focused on the plan I would leave my daughters that I was in danger of forgetting what tied the present and the future together. How I learned to ground myself in uncertainty would help me face my own death, a planet in crisis, and my daughters' lives.

After our visit, Trish sent me the names of several death doulas, ranging in age from their forties to their seventies. In the days that followed, I met with each one for visits in coffee shops around town. Among the fifty people trained as an end-of-life doula in this area, the three she wanted me to meet, I would later discover, would have some thread of connection to my parents. In this journey, the surprises that first blew me away became expectations, like a shape I could see in the clouds, if only I were looking.

When I met Carole Shoaf at the Pennycup Coffee House, she first shared her discovery that good friends of hers had known my parents through their hikes on the Appalachian

Trail. She'd realized the connection only after reading an essay I'd written about their deaths.

With her soft gray curls and a youthful face that seemed relaxed but radiant, this violin teacher and death doula leaned in to speak as if we were old friends. She talked about her own family and how she became an end-of-life doula. "My parents had a study group that met weekly to read about Rudolf Steiner's views on death," she said. "Steiner believed the soul leaves the body gradually over three days, and my mother and father started a cooperative to support each other in holding vigils when they died."

"So your parents were the ones who influenced your feelings about death?" I asked.

"Yes, you can see why I wanted to meet you," she said. "It wasn't just about death and dying. It was about our parents as mentors guiding us through death."

Carole described her experience of holding a vigil for the deaths of both her parents and drawing on the members of their cooperative for support. For her father's death, the women washed the body and put him in a cardboard casket; they used dry ice to keep the body cool; she played the violin for him, and they had a "sing out" on the third day, when the funeral home picked up his body to be cremated.

"We need to give the funeral homes credit for the work they can do," she said. "They were so respectful."

About a year and a half later, her mother looked at her and asked, "Do I need to eat anymore?"

"We won't force you to eat at all," Carole said. Her mother went into a coma that same day and died one week after her request.

Surrounded by the sounds of baristas brewing coffee and washing dishes, I cupped my hand over my ear to hear as she described her beliefs about the dead.

"I believe that the living and the dead are separated by only a thin veil," she said. "Much as we get inspiration from looking at art, the dead gain from our spiritual thoughts, and they are always helping us. They are so accessible, and they need our spiritual growth."

To honor that crossing is one of the most important events we can witness, she said.

"I want to work with the dying as a way to honor what my parents gave me," she said.

A moment later, she added, "Because of them, I don't fear death."

The steam of an espresso machine hissed loudly behind me.

"Shit," I whispered. "That's exactly how I feel."

She nodded as if she already knew.

"Well, I wonder what connection I'll have to this next death doula?" I asked Lyn during our daily chat on the phone.

"At this point, you know there will be something," she said. "It's just a matter of how long it will take to discover!"

"Did I tell you about the death doula I met from Greensboro, who lived down the street from my mother's first cousin?" I asked.

"Nothing surprises me at this point," she said.

"This woman brought a handwritten list of resources for me," I said, reaching for my notes. "Rudolf Steiner's *Staying Connected*, Stephen Jenkinson's *Die Wise*, Long and Lynch's *The Good Funeral*. And here's the crazy part: her handwriting looked exactly like my mother's, perfectly compact with an elegant slant to the right."

"I've been meaning to ask," Lyn jumped in. "How are you going to have a home funeral in a campus rental? Are you going to ask permission from Human Resources? Do you think the college would allow it?"

"Why couldn't I have my body in a rental?" I asked. "Your cat drags dead animals into your house on campus on a weekly basis. But I probably won't be in this duplex when I die. I can't stay in the house after I retire—well, if I can ever afford to retire."

"Your parents died early, so you never know," Lyn said. "But you don't even have space in your bedroom for a massage table to put your body on! Do you want your body leaking on your bed?"

"I hadn't thought about it," I said. "but you're right. I need a new mattress, so maybe it wouldn't matter. In the old days, people just laid on their beds. That's what my dad wanted."

"I have a perfect idea!" Lyn said. "You've got that expandable coffee table in your living room that unfolds to make a dining table. You could just put your body right there in the middle of the house!"

In our 900-square-foot rental, we had one combined living room and kitchen and two small bedrooms. So Lyn's suggestion—made in jest—was to put the body in the middle of the action, using my coffee table designed for small spaces.

"People could eat their casseroles and pies right by you and do all those rituals you love," she said. "I mean, if you're going to plan, you might as well assess all the options."

"I think my children would sacrifice the mattress, rather than keep me that close to the kitchen!" I said. "That would send Annie Sky straight to the Motel 6!"

I wanted to talk to someone who had planned a home funeral with the help of the Center for End of Life Transitions. A friend from church, Tahani, had buried her father and her mother on their property about twenty miles from Asheville, where it was legal to do on private land outside the city limits. While still in good health, Tahani's father had contacted the

Center for End of Life Transitions. One of their death midwives, named Ruth, came to their house and talked to the family about home funerals and natural burial. Five months after that meeting, her father collapsed at his desk and died. In that moment, Tahani called 911. Firefighters came and attempted to revive him, but he was dead.

"I asked them to move him onto the bed in the guest room," Tahani said. "I didn't really know what to do next, but I happened to see Ruth's card sitting on the coffee table."

After she called, Ruth came to the house, helped her wash her father's body and dress him in a black shirt and a tiger-skin rami, a colorful cloth worn around the waist, reflecting her father's adventurous life.

"Ruth instructed us in the placement and changing of the dry ice and explained what we might expect over the next few days," she said. While Tahani's extended family traveled to Asheville, they decorated the room with autumn leaves and candles and played music her father would have liked.

He died on a Saturday, and they buried him on Tuesday.

"I had never seen a dead person before," she said in her lilting accent, a product of a childhood spent in Libya, Australia, and Louisiana. "I remember thinking how severe but dignified my father's face looked, and I was struck when my mother died that she had the same look. They looked like themselves but not alive."

Tahani's three children played key roles in preparing the burial site about a third of a mile from the house. The family dug the grave, and her husband converted a beehive stand to a stretcher with removable carrying handles.

"My daughter Miranda served as the body in our dry runs," Tahani said. "My father was very good at logistics, and we vacillated between congratulating ourselves at arriving at some solution worthy of him and thinking how he would sneer at our mistakes."

This sounded quite familiar to me as I recalled my siblings and I discussing logistics in the hours after my father's death, hoping we'd met his expectations.

They used a bedspread as a shroud. Since it was November, the family gathered colorful leaves to throw on the body when he was in the grave, before covering him with soil.

"It was hard to do that," Tahani told me. "No one wanted to be the first to cover his head."

They found a stone nearby for a headstone, and her brother, a blacksmith, created a plaque with his name and dates of his birth and death.

When her mother died next, Tahani again called the center, and Caroline picked up dry ice on her way to the house to help. After our priest, Milly, prayed the last rites, she described those rituals as more for the survivors than for the dying.

The logistics of this burial were similar to her father's, although her children were older this time. "It will be very different for them, I imagine, when they are faced with the deaths of their parents and partners, to have had this experience," she said.

"We chose a lovely exotic cloth for my mother's shroud," she said. "It was something to watch my daughter and her cousin sewing it up around her body."

She told me that when they carried her body to the gravesite, they startled a great blue heron, which flew up and gazed from a tree above. The family considered it a sign, which they didn't have to understand to cherish.

If I closed my eyes, I could imagine placing a drop of sandalwood oil on the forehead of my mother, who smelled of Jean Naté cologne when I was a child, Chanel No. 5 as I became a teenager, and Sleepytime tea mixed with smoke from the wood stove when I was an adult. She would have loved the ritual of anointing oils, in life or death.

My daughter Maya sometimes spoke about wanting to be a "radiating spirit" to those she mentored as a camp counselor in the summers. She could have learned how from my mother, whose radiating energy must have touched my father's heart when he held her body close to him.

This is what I choose to believe. These rituals of spending time with the dead might be for the living—for radiant living—to connect us to our relationship with the dying. Preparing my father's body brought me closer to that liminal state between life and death, whatever it may be, and filled my senses for a lifetime of recollection.

I'd begun to carry a rather simplistic view of what happens to the soul, believing the dead become a part of the love in our hearts, a direct line to protection and resurrection here on earth. My next steps to find a final resting place weren't only about my death or my children's lives in a world upended by climate disruption. They were about creating a new heaven and new earth—here and now—that can sustain us all.

Chapter 5

WHEN DEATH PROTECTS THE LAND WE LOVE

Natural Burial in Conservation Cemeteries

"WHAT ARE YOU GOING to wear to volunteer at a burial?" Lyn asked me when I stopped by her house on the way home from work.

It was a valid question, one I hadn't considered.

"I have no idea," I said. "In photos on their Facebook page, the director, Caroline Yongue, wears flowing clothes with neutral colors—kind of like Chico's meets the monastery."

"Let me know if you need to borrow something," she said, pulling her loose wrap over her shoulders with a theatrical flair.

About a month earlier, I'd put my name on the volunteer list for Carolina Memorial Sanctuary, the only conservation burial ground in North Carolina, located about a half-hour drive from my house. The week of the home vigil I'd attended, I also received an e-mail from Cassie Barrett, manager of operations and marketing at the burial ground.

"Are you available to volunteer at a burial at 1:00 p.m. Friday?"

I didn't hesitate. "I'll be there!" I wrote back within seconds.

Cassie had shared a document she sent to volunteers, a version of "what to expect when you're expecting a natural burial." She'd told me the burial would be for Paula, mother-in-law of Aditi, the Indian American hospice and palliative-trained physician. By chance, I'd received two separate invitations—one to attend Paula's home funeral and the other to volunteer at her burial.

I'd walked the grounds at Carolina Memorial Sanctuary but never witnessed a burial at a conservation cemetery that protects the land in perpetuity from development. Volunteering seemed like a practical way to give back while evaluating if this might be the right end-of-life choice for me. The cemetery had partnered with Conserving Carolina, a land trust that held the voluntary legal agreement for land protection, also called a conservation easement. This meant that those who chose to be buried there were doing their part to invest in land conservation for the long haul. The easement limited future use of the property to maintain its conservation value.

Located only a few miles past the Sierra Nevada Brewery near the Asheville airport, this burial site is one of sixteen conservation cemeteries in the United States. Six are located in the South—in Florida, Georgia, Tennessee, North Carolina, and South Carolina. The others are in Ohio, California, Illinois, New York, Oregon, Washington State, and Vermont. Sixteen burial grounds may not seem like many, but interest in this type of conservation is growing as people learn how the end of life can also address climate change and habitat loss. Additional evidence of this interest comes from the National Funeral Directors Association, which found that 72 percent of US cemeteries report increased demand for green burial.

At its simplest level, natural burial is defined as burial of an unembalmed body in a biodegradable container without a vault. People often use the terms *green burial* and *natural burial* interchangeably, and this type of disposition can occur in many places—conservation burial grounds, conventional cemeteries, and even a backyard.

I'd first visited the conservation burial site in North Carolina about a year earlier, as I prepared to give a presentation at my church about green burials. The story line made good sense for me: bury an environmentalist at a conservation cemetery. I knew my parents would have approved, although I still didn't know what my daughters would think. At the start of my research, I believed my options were simply burial or cremation and didn't realize the diverse choices available until I encountered them along the way.

Green burial doesn't carry the carbon footprint of cremation or the excessive waste and pollution of conventional burial with gallons of embalming fluid buried in the ground. I'd listened to my father preach against the practice of embalming since I was a child. But I'd read about a survey by the National Funeral Directors Association, in which only 48 percent of respondents were aware that embalming isn't necessary.

Other common characteristics of natural burial sites include the depth of the grave, which is about three to four feet, rather than six feet, to promote decomposition of the body as more oxygen and bacteria are present at that depth. The gravesite also uses native plants and trees, and if headstones are used at all, they are often natural stones native to the burial ground, lying flat.

These practices differ somewhat from my father's burial: In our neighborhood cemetery in Alabama, the gravedigger dug the hole to the typical six feet, and we used a headstone that wasn't made of native stones. My dad had even quoted the price in his directives to us, as he wanted a marker similar to my mother's.

Although we didn't know it at the time, the Green Burial Council was founded the same year as my father's death in 2005—as an organization that certifies three categories of green cemeteries based on specific criteria. Some conventional cemeteries, called hybrid burial grounds, offer the option of burial without a vault and don't require embalming. The second type, described as natural burial grounds, prohibits vaults or burial of embalmed bodies or burial containers that aren't made from natural materials. Carolina Memorial Sanctuary is an example of the third kind, conservation burial grounds, which meet the requirements for natural burial and *also* protect land in perpetuity. I'd first learned about the differences from the Green Burial Council's website, which provides a map of cemeteries that meet the different standards.

Since my dad's death, I'd talked to the founders of the 112-acre Larkspur Conservation Burial Ground outside Nashville, adjacent to 172 acres of protected land owned by the Nature Conservancy, and visited other conservation cemeteries. But I now had the opportunity to volunteer at a burial site that hadn't existed when I originally wrote my final wishes for cremation of my body. Once I had a chance to peruse the orientation materials for volunteers, I called Lyn to report on the dress code.

"We're supposed to wear neutrals, not somber tones like black," I said. "I might end up directing traffic, and I can view the burial with the staff and family."

"My closet is your closet," she said.

"Did you know that conventional cemeteries function like landfills?" I asked, moving into teacher mode.

"Is that so?" she said. "Do tell."

"Think about a ten-acre cemetery like the one where your mom and dad's ashes are buried," I said. "The ground contains enough coffin wood to construct forty homes and another 20,000 tons of concrete from vaults."

"That's all underground?" Lyn asked. "But the graveyards look so peaceful." She laughed and added, "With those stats, you should come with us to Trivia Night sometime."

"The embalming fluid in those ten acres is enough to fill a swimming pool!" I continued. "Those chemicals leach into the earth and pollute the groundwater, not to mention the pesticides used to maintain the grass."

"Well, that's one reason I chose cremation, but now you're making me rethink that decision too!" Lyn said. As we wrapped up our call, she said, "Let me know how the volunteer shift goes—and what you decide to wear."

I continued to mull over those stats and implications. Even in the 1970s, my father had been concerned about these environmental costs of modern burial. He also wanted to avoid the exorbitant costs of death: the average price of a green burial is $5,000, as opposed to $10,000 for a conventional one. My parents had purchased their plots in the local cemetery years before they died. Funds spent by families on expensive burials meant less money for the living. After my father's passing, I knew that natural burial connects family and friends to death in an intimate way. The environmental and financial impacts were only one part of the story.

Standing in front of the full-length mirror in my bedroom, I chose my unorthodox funeral clothes: a loose white shirt, tan knit skirt, and comfortable shoes for walking over uneven surfaces, as advised in the volunteer handout. My breezy outfit contrasted with the more conventional dark navy clothes my aunt kept in the trunk of her car for years, in case an aging relative died during a visit to my grandparents' house in Mississippi.

I drove to Carolina Memorial Sanctuary, which was next to the Unity church that sold the land in 2015 to create this innovative yet ancient type of burial ground. Within minutes of arriving, I saw Cassie and Caroline, who both worked at the sanctuary, pull up in a golf cart. They introduced me to Steven, the burial and land stewardship director, and Matt, a bearded gravedigger who published a zine about natural burial. We were the team of staff and volunteers for the burial that afternoon.

Cassie explained my role as a volunteer parking attendant, which involved the basic task of directing cars to their proper place. When the first vehicle arrived, I ran to my official spot, raising my right arm with authority and pointing to the small gravel lot. I tried to smile but not too much, look serious but not too somber.

The night before, when I'd attended Paula's home vigil, the family tenderly cared for her body and grieved with friends around them. Those who came to the burial brought that same compassionate care to this outdoor setting. While I directed vehicles, Aditi and her family passed in a black minivan carrying Paula's body wrapped in fabric in the covered cargo bed. Down the road, Cassie waited to help them place the shrouded body, covered in sunflowers, daisies, and other wildflowers, inside an open woven carrier with handles on the side and a firm surface on the bottom.

Soon the golf cart and trailer with the basket approached the lower parking lot, much as a landscaping crew would carry tools to a site. Like an exquisite sunset or a horrific natural disaster, her shrouded body was a magnet; I couldn't look away, even though I was supposed to keep my eye on the road for cars.

How we leave the earth is not so different, I thought, *despite the differences in how we die.* At that moment, I was alive, but in the future, my body would be held similarly in a shroud, an urn, or a casket, much as my own parents' bodies had been. In that place, it was easy to imagine the people I loved walking slowly along a wooded trail with my body. The volunteer stint felt like a dress rehearsal for the end, in much the way that holding a friend's newborn baby provided practice for having my own.

The service was scheduled to start at two o'clock, and Cassie gathered the fifty people—from toddlers to seniors—around the golf cart and trailer. While we waited, six children hopped into the back of another golf cart and pretended to steer on an imaginary road trip.

An elderly woman asked me in a whisper, "Now, are y'all gonna be with us at the site? Is the hole already dug? Will we cover the body with dirt?"

"We'll be with you the entire time," I said, shading the bright sun with my hand. "The hole is dug, and everyone will have a chance to close the grave together."

Dressed in a loose white tunic, Cassie got everyone's attention, and we processed behind the golf cart along a gravel path in the woods. A man with a cane asked me to point out grave sites to him, which I did by noting the raised earth, often hidden behind grass and trees. "Each site is marked by GPS," I told him.

Walking in a conservation burial ground feels completely different from visiting a landscaped cemetery, as the graves integrate into the woodlands, in this case, with successional plants and trees such as tulip poplar and white pine that typically grow after an area has been cleared. We walked on a trail in the woods, as opposed to sidestepping headstones in grass.

"So life just grows up around the grave," another man said. Assuming the role of official volunteer, I explained that if he wanted to easily visit the grave site, the staff would clear a path, given three days' notice. And anyone could walk the designated trails at the burial ground at any time. He nodded, somehow reassured.

In the hot sun, we soon arrived at the site, where a hole had been dug three feet deep. I could see three separate piles of dirt on top of tarps, to ensure the soil returns to its original layer, with the topsoil placed back on top. So the first tarp had soil from the bottom of the grave, along with rocks that they pulled out of the dirt. There were pine boughs inside the grave and at its base to provide a buffer between the earth and the body. Caroline pulled out six rattan folding chairs from the trailer, but most people found a place to sit on the ground or stand by the grave site. The children gathered at the feet of Paula's body in the wicker carrier.

Without words, her grandchildren began pulling pine needles off the boughs and placing them around the body, almost like natural adornments. Earlier, they'd painted stones with bright colors, which they planned to put in the grave.

Sitting behind them, Aditi began to play a harmonium, a small organ-like instrument, and her husband, Jay, strummed his guitar. A friend of the family who was a chaplain officiated the informal service, calling on different people to read prayers.

During the offerings, a three-year-old boy called out, "I'm hungry," and Aditi's mom dug into her bag for crackers and nuts. She'd come prepared, especially as we'd been told the service would last about two hours.

After about an hour, the time came to lower the body into the grave. Cassie explained the step-by-step process of lifting the body by the sheet in the woven basket.

"Could I have five volunteers?" she asked.

Five men quickly raised their hands. Cassie stepped into the grave, three feet deep, and asked two of the men to join her. They lifted the body off the gray sheet and placed her on the pine boughs in the grave.

"We ask everyone to take a pink rose and place it on the grave," said one of Paula's six sisters, carrying a basket of flowers. "You can share one word or phrase about her if you'd like."

When the shrouded body was covered with roses, Cassie picked up a shovel and shared the tradition of closing the grave. "A Jewish family introduced us to this way of shoveling the dirt, and it's become a ritual for us," she said. She picked up the shovel, turning it face down: "For your first scoop of dirt, you turn the shovel face down, which signifies your reluctance to let the person go. For the second and third scoops, you shovel the dirt face up, which signifies your willingness to let them go." She demonstrated with three scoops and then asked for volunteers to follow her.

She lifted a pint-size shovel, and the children came forward to help. I recalled my daughter Maya shoveling dirt into my father's grave when she was only six years old. At least twenty people scooped soil following the Jewish tradition, while others picked up fistfuls of dirt, kissed the dirt, and threw it into the hole. Then it was time to close the grave.

"It's easy to get into a rhythm of shoveling and forget that others may want a turn," Cassie said. "So just be mindful."

The hole had looked cavernous and daunting to me: I'd thought it would take an hour to close. A friend who volunteered at the sanctuary told me it took four hours for two people to dig a grave.

But within fifteen minutes, the grave was sealed with soil. The staff at the sanctuary ensured that everyone added the earth in the same layers as it had been dug. They then lifted the tarp so the rest of the dirt could fall into the grave.

The burial site was mounded several feet higher than the rest of the earth. Thus, the soil could settle when rains came, and the grave wouldn't sink when the body decomposed.

Cassie picked up bushels of pine straw and showed us how to cover the mound with the straw. She also had small sticks about a yard in length and would stick them into the dirt when it was level with the ground.

"What are those sticks for?" a woman asked me. "Is that to keep out critters?"

I walked over to ask Cassie, who said the sticks would help aerate the soil, so new trees and plants could establish themselves, and also would introduce beneficial microorganisms.

The last touch was to place the fresh flowers from the clear vase on top and encircle the grave with rocks dug from the earth, creating a visual border. Aditi's brother took pictures of the grave and waited until everyone left to take a final shot of this resting place. Throughout the two hours, people cried, laughed, sang, soothed children, hugged each other.

"Have you ever been to a burial like this before?" one of Paula's sisters asked Aditi's brother.

"No, never," he said.

"It's been therapeutic for us all," she said. "It's exactly what Paula would have wanted."

The vision Caroline Yongue pursued in her quest to create a conservation burial ground in Western North Carolina was this: honoring the land with inclusive traditions to honor the dead. For thirty years, she'd worked as a hairdresser, listening to the details of her clients' lives while she also pursued a spiritual path of meditation in community.

Now in her early sixties, Caroline exuded the presence of someone comfortable in her own body. During our conversations that year, she'd told me that when a person dies, the elements in the body dissolve: "It's helpful to consider what will allow you to leave the body," she'd said. I'd been so consumed with how to dispose of my remains that I hadn't considered an exit strategy for my spirit. In my research as I interviewed death doulas, they spoke about the spirit gradually leaving the body in the three days after death. And I'd wondered about the impact of my parents' sudden deaths on the departure of their souls.

As she worked with the dying, Caroline saw the need for a conservation burial ground to serve people who want a closer relationship with death, community, and the land. She found thirty acres in the nearby town of Black Mountain, but then she was diagnosed with stage 3 adrenal failure, her marriage fell apart, and the economy tanked, a triple-bottom collapse in her life. So she put the vision on hold even as she continued to lead workshops on death and dying and became an ordained Buddhist monk.

Unexpectantly, she was approached by leaders from the Unity church, who expressed interest in selling eleven acres to start a conservation cemetery. She made a bold move by putting her house on the market and giving the proceeds to her sangha so they could buy the property in 2015. So that others could die in a way that conserves the land, she gave up her home.

"What did I have to lose?" she asked me. (*Ah, everything?* I thought, trying to look Zen.)

Caroline said her own siblings and parents didn't understand the decision, but she never looked back. She worked as a caretaker at her Buddhist sangha, which allowed her to live on the premises and use the office space for the Center for End of Life Transitions. As a teacher, I'm sometimes asked by my students if one person can make a difference. Looking at Caroline's life, I saw the answer.

That same summer, I returned to Carolina Memorial Sanctuary as a potential client or "guest," as manager Cassie Barrett calls visitors interested in the site as a final resting place. This burial ground had joined with other conservation cemeteries across the country to form the Conservation Burial Alliance, promoting this form of stewardship of land and community. During our meeting, I shared the questions at the heart of this journey: Is this the place? Are these the people? And also, as my father's child, is this the price? I was at the beginning of the search for a sustainable end of life, but I'd been drawn to conservation burial grounds ever since I discovered them years ago in Georgia and Florida. Now that possibility existed close to home.

As Cassie was beginning the tour, I hopped into the golf cart with her. As a Warren Wilson College graduate, she'd found a job where she could work with people and the land—and wear tan overalls with quilted patches and a button-up white- and blue-striped shirt. For so many of my students, this seems like a dream job. Later that year, one of my own students would be hired as a land steward for the cemetery.

"An important part of the narrative is our relationship with death in this culture," she said. "People often feel uncomfortable with death, but we each have a responsibility to tell our loved ones what we want done with our body after we die. We're holding space for people to encounter death in community."

These principles weren't hypothetical to her: Cassie's stepmother had been buried at the sanctuary after she died of terminal breast cancer. That diagnosis prompted weekly conversations about planning for death in her family.

As the seasons changed, the cemetery changed as well. Spring and summer brought green grasses and vibrant wildflowers, in full display during my visit. In an open meadow, we paused at the Pet Memorial Garden. "We are a whole-family cemetery," she said. "If someone has a spot for themselves, they can have their pet buried next to them."

The staff at the land trust Conserving Carolina decided to pursue the conservation easement that would protect the land from development due to the ecological value of McDowell Creek and the wetlands on the property. To that end, the sanctuary received a matching grant for $142,000 to restore the creek and the wetland as a rare mountain bog.

"This property was a dairy farm at one point," Cassie said. "You can see where they filled in the wetland for agriculture. To restore the land, we'll dredge the earth, remove the invasive species, and repopulate with native species."

In terms of the price, the Sanctuary includes three areas where people can buy plots: Woodland ($3,500), Meadow ($4,000), and Creekside ($4,500). The funds go toward maintenance and ecological restoration of the land, as well as opening and closing of the grave. As at most cemeteries, this cost doesn't include the services of a funeral home, which many families don't use, as they can transport the body themselves to the site. In other cases, a funeral home might provide only transport, which was reflected in a lower overall price.

We passed several sites that had been recently dug, each with three piles of soil on tarps, waiting for the closing of the grave. A wooden board and a tarp covered the hole, so animals wouldn't fall in and get caught. At the time of my tour, they'd completed fifty-six human burials and nineteen pet burials and had sold 330 plots total. In consultation with other conservation cemeteries, the sanctuary put a cap of 3,000 bodies on the land, with no constraints on the burial of cremated remains.

As the golf cart pulled back to our starting point at the parking lot, I marveled at the options so close to my home and across the country—institutions embracing death and dying in a way that honors the places where we live. When I started this research, I didn't realize the wealth of people working to bring a good death closer to my life. As an environmental educator, I saw the benefits of a conservation cemetery for my family and community, but I still wondered: How could my burial on the eleven acres at Carolina Memorial Sanctuary make even a small difference with the changing climate?

Many of the other conservation burial grounds, such as the ones in Florida and Tennessee, are adjacent to larger tracts of protected lands. I'd read a study in the journal *Urban Forestry and Urban Greening* that natural burial can contribute to a range of ecosystem services such as mitigating flooding, loss of biodiversity, and poor air quality. City planners even see conservation burial grounds as one tool for creating open spaces for people and promoting biodiversity in urban spaces. This made sense to me, as cemeteries once provided green areas for recreation and even picnicking in cities across the country.

Carolina Memorial Sanctuary began collaborating with ecologist Shaun Moore to start the process of restoring the creek and the wetlands. In the smallest of worlds, Shaun and his wife had lived for nearly a decade as our neighbors at Warren Wilson College, and students from the college were helping in the restoration work. He'd examined aerial photos of the property from 1951, when the tilled site was bare but for a few trees. Before the land was cleared for farming, the property included six acres of wetland, as revealed by soil samples.

"When people purchase a burial plot here, they're investing in the long-term protection of habitat for wildlife and green spaces for humans," Shaun told me.

Supporting conservation is exactly the kind of legacy I'd like to leave for my daughters.

When I looked at the ecological research, I found a 2017 study that compiled a database of known natural burial grounds in the United States. Of the 162 sites, 99 are hybrids, 54 are natural burial grounds, and 10 are conservation burial grounds with the commitment to restoring landscapes, enhancing ecological services, and protecting land through easements. The authors conclude, "The US practice of burying embalmed human remains in a lawnpark cemetery is not environmentally sustainable." A 2021 inventory revealed a total of 300 green burial cemeteries in the U.S., including hybrid, natural and conservation burial grounds. The scientific evidence supported what I'd seen from volunteering on the ground: this model—burying bodies in a sustainable way to conserve land and the species around us—can be replicated. And I could have a small part in the practice.

After my father's death, when I first started to consider the impacts of natural burial on climate change, I turned to Susan Marynowski, a friend from graduate school, as well as an herbalist, environmental educator, and a founding board member of the 93-acre Prairie Creek Conservation Cemetery in Gainesville, Florida. This burial ground, with meadows and live-oak hammocks, sits adjacent to Alachua Conservation

Trust's 512-acre Prairie Creek Preserve and the 21,000-acre Paynes Prairie State Park. The partnership of the cemetery with the local land trust created contiguous protected areas for conservation. During the past few years, we have talked about the intersection of the climate crisis and the burial of humans, two issues that many people don't connect.

"I think of all that the earth has given me in my life. I've taken food, water, land, energy," she said. "And in my death, I'm going to return to the earth. A wolf dies, and its body lies in the woods, and a vulture might eat it. We can put our bodies back into the ground. It's like composting in spirit; all of the molecules go back into circulation."

She believes people need to band together to create community burial grounds protected from development.

"Conservation burial is a way for dying humans to have a positive impact on the natural world," she said. "This is something that communities can do together to preserve and restore land, provide beautiful burial places, and make space for meaningful ceremony."

She emphasized that these efforts can involve both conservation cemeteries and conventional cemeteries. "The small-town cemeteries nearby don't require vaults," she said. "Vaults are a product of what are called green-lawn cemeteries, where people want the manicured land to be level."

Prairie Creek Conservation Cemetery works with any of the local funeral homes, including a low-cost mortuary that provides less expensive options for storing and transport of the body. A group of five to ten volunteers, proud members of a grave-digging league called The Society of the Pick and Spade, digs each grave.

"They marched in the University of Florida homecoming parade, wearing top hats, brandishing shovels, and pulling our Amish burial cart," Susan said. "We were playing the Tigers, so a tiger kept popping out of the casket!" With that punchline, Susan laughed her full-throated chuckle, her authentic mix of humor and earnestness, which I first loved twenty years ago in school.

She described the individual expression encouraged at these burial grounds. People can make their service what they want it to be.

"At Prairie Creek, we've had a Muslim burial, a military burial with honors, and for a person who worked with service animals, a burial with fifty dogs around the grave," she said, laughing out loud at that image. "Nearly every family has helped to fill the grave. It is a great catharsis and closure for people to help with the burial—especially the kids, as that's what they will remember."

That was certainly true of my daughter and what she has remembered of my father's funeral.

As we spoke, I felt again the need to act—much like the urgency of youth-led protests for structural climate action—to make informed decisions about my individual life and death, even if I am just one person. With my students, I advocated for public action to pressure governments and fossil fuel industries, and I know that our collective voices matter: But I was the only one who could make decisions—right now—about my body and the end of my life.

With each new conversation and piece of information, I mentally played with hypothetical scenarios. Imagining my end, I envisioned my daughters shoveling dirt into my grave in the natural setting at Carolina Memorial Sanctuary. While volunteering at the sanctuary, I viewed this land as holy ground. I knew the $3,500 for a plot would be a significant investment—

one to save for over time. I was also aware it was much less than the $10,000 average cost of conventional burial, especially if we kept my body at home and saved on expenses from a funeral home. The conservation cemetery seemed like an obvious choice. Yet I was ready to examine the options for burial even closer to home, on the campus where I'd raised my children.

Perhaps too eager for her input, I took my daughter Annie Sky to visit the sanctuary that fall. I knew it didn't make sense to rely on a fourteen-year-old for my end-of-life plans, but I also realized my children might be my primary caretakers and would share their opinions until the end.

Upon arrival, I parked in the lot and began to walk with her on the trail.

"Where is the cemetery?" Annie Sky asked.

"This is it," I said.

"Whoa, is that a body under there?" she asked, pointing to a mound of dirt covered with dried flower petals. We walked through the woods, slowly acclimating to the presence of grave sites, hidden by bright colors of the fall leaves. A white-breasted nuthatch flew above us. Squirrels ran up and down tree trunks. The blue sky was crystal clear, life beginning anew.

"Well, this is pretty," she said. "I'll give you that."

I nodded, energized by sharing this place with her.

"But honestly, I don't know if Maya and I would drive this far out to come visit you," she added. "So you might want to consider that fact—I mean, if that matters to you at all."

In that heartbeat, I imagined my own mother smiling at the radical honesty of teenage daughters, facing life and death together.

Chapter 6

BURY ME CLOSE TO HOME

Green Burial in Conventional Cemeteries

My daughter Maya was a first-grader when my father died. At twenty-one years old, she doesn't remember walking into the standing-room-only church as the organ played the hymn "All Things Bright and Beautiful." She can't recall the sounds of my dad's bluegrass band when they sang "When the Roll Is Called up Yonder" at the cemetery with its ancient oak trees.

"The only thing I remember," she told me, "is shoveling dirt into Grandad's grave."

In the heat of that June afternoon, fifteen years ago, I watched the children bend their tender, lean bodies to peer into the grave while clumps of earth fell on solid pine.

As a young adult, my eldest now collects remembrances like exquisite seashells, posting quips to her friends on Snapchat: "Let's make memories this year!" So I viewed her recollection of my father's burial as something precious. Sitting in my kitchen, she could describe concrete details like taking turns holding the shovel with her cousin as the soil landed on the casket with a deep and mighty thud, like thunder underground.

During our annual pilgrimage to the small cemetery in Fairhope, Alabama, we would stand by my mother and father's headstones and recite aloud the Lord's Prayer. It was usually hot and muggy in the summertime; the children were sunburned and impatient, racing through the prayer to return to the air-conditioned car and the promise of sleep on the long drive back to North Carolina.

At the graves, there was a slight dip in the ground where the caskets and my parents' bodies had decomposed, side by side, returning to the earth. The grass resembled the edges of a cake that hadn't completely risen. No one seemed to mind. The land looked more rugged, less manicured, much like their lives. Around the headstones, fire ants built their nests, which my daughters sometimes knocked down with a stick, although the biting insects would begin to reconstruct their homes before we even departed.

"People tend to die much like they've lived," a death doula and nurse once told me.

My father didn't want his death to pollute the ground he'd traversed with my mother, hiking with red-clay bricks in their backpacks from their house to the cemetery to train for their long-distance treks. We didn't know it then, but much as humans buried their dead for ages, we were practicing what's now called green or natural burial.

In his directives, he instructed us to handle his body in the same tradition as many Indigenous cultures, as well as Jewish and Muslim faith traditions. "For my funeral, the plan is that the funeral home tasks, which primarily involve transportation, would be done by family, friends, and the church," he wrote.

Green burial can happen in a conservation burial ground, a local cemetery that allows burial without a vault, and even private land. Since I didn't own my house on campus, that choice wasn't an option for me. But many neighborhood cemeteries allow green burial, as long as the contract doesn't require a vault.

At the beginning of this project, I didn't recognize my own access to a cemetery on the campus where I'd lived and worked for twenty years. Since starting my job at Warren Wilson College in my midthirties, I'd heard about the resting ground where the college founders were buried on the thousand-acre campus—and then I'd forgotten about the landmark in the rush of my days. I didn't know my home as well as I thought—or the obstacles to green burial on the sustainable campus where I lived.

Between the two options, cremation is more sustainable than burial of an embalmed body with a concrete vault underground, but a natural burial allows life to continue after death. In many ways, green burial was also a way to talk about resurrection, since the body returns to the earth to nourish other life.

That is the language of my faith tradition: Resurrection. Life. The Episcopal *Book of Common Prayer* speaks about death with these words: life is changed, not ended. My mother died when she was close to my age now. *"Punto final,"* as my oldest says to me in Spanish: That's the end. But on the journey of this year, I'd begin to walk among the graves of people who built the college where I lived—the place where I'd raised my two daughters—and I would also feel the actual presence of those who came before me.

One muggy summer evening, I ran into my neighbor Susan, walking her two dogs, a black and white border collie next to a beagle without a functional front leg, who hopped to get around.

Trained as an economist, Susan knew a thing or two about facing death, as she'd gone through chemo, radiation, and immunotherapy and was now cancer-free. That night, we both marveled—with a note of fear—at the scorching heat that summer, the hottest July on record in the face of warming global temperatures. When I told her about my research, she shared her intentions to be cremated, although she wasn't certain if she'd written down her final wishes. (I was learning that if folks weren't sure, they often hadn't documented the details.)

"I don't care what my girls do with my ashes," she said. "If they throw them into the Swannanoa River, that would be fine with me." We'd raised our daughters together on this campus, where they ran back and forth between our houses, returning home when Susan rang a cowbell outside her kitchen window.

"But you know," she added almost as an afterthought, "I also own two plots in the Warren Wilson Cemetery." She'd bought them for $75 each with her former husband and received the deeds in the divorce settlement.

"I can put two cremated bodies in each plot or one body each," she said. "Hey, I could sell one plot to you, and our girls could come visit us there together!" We cracked ourselves up at the thought of lying side by side on campus, much as we'd raised our children.

"I can't believe I didn't know we had the option of purchasing a plot at Warren Wilson," I said. "I could be buried in a local cemetery near my house, just like my parents." In my readings, I'd learned that the word *cemetery* comes from the

Greek word *koimeterion* and means "a sleeping place." It made sense to rest—and decompose—close to home.

The next day, I drove to the burial ground—or what I thought was the site, but I ended up at the cemetery for a nearby church. Lost in my own neighborhood, I resorted to an online search for "Warren Wilson Cemetery" to orient me. Google Maps revealed its location right around the corner from that church. The week before, while driving my younger daughter to a babysitting job, I'd passed the sign for the college burial ground without seeing it.

When I called Lyn that night, she told me the Presbyterian church on campus owned and managed the site. That night, I emailed Rev. Dr. Steve Runholt to get information about the possibility of securing a burial plot. He replied the next day: "It's the one piece of property on campus that the church owns outright," he wrote. "Our three cemetery trustees maintain the grounds and arrange for all the interments."

I'd known Steve for his impassioned activism—from immigration reform to climate justice—and for his role as pastor at the Warren Wilson Chapel, the imposing chalet-like church in the middle of an increasingly "spiritual but not religious" campus.

The day after writing him, I popped into his office to talk about death.

"When I first started at the chapel in 2005," Steve said, "I got up to do the call to worship, and I looked out at the parishioners and realized many of them would die soon. It was an aging congregation."

In his church, Steve presides over about five or six funerals a year, but almost everyone who died in the recent past had been cremated. If families asked, he recommended the low-cost Asheville Area Alternatives, whom I'd interviewed earlier in the summer, because they provided cremation at an affordable price.

"It's been five years since I've done a burial of a body," he said. "Most everyone wants to be cremated."

"I thought I wanted cremation as well," I said. "It seems so convenient! But I've been concerned about the fossil fuels burned in the process, when we have other options like green burial closer to home."

Much as I'd posed questions at potlucks and book clubs that year, I asked Steve about his own final wishes for his body after death. He put his hand to his head, running his fingers through his hair, and leaned back in his chair as if contemplating the options.

"You've touched on an important point," he said with a sheepish nod. "I know exactly what I want, but I haven't written anything down." He wanted to be cremated with his ashes spread in places of significance to him, such as the Warren Wilson Cemetery, the Black Hills of South Dakota, and the garden at Wadham College in Oxford.

I admitted my initial ignorance about the location of the Warren Wilson Cemetery and my desire to learn more.

"It's always a place where I feel like I have to take my shoes off," Steve said, getting tears in his eyes as he recalled the community members buried there.

"Were you there for Callum's funeral?" he asked me.

I'd gone to the memorial service for the tragic death of our dear friend Callum, an adventurous red-haired young man, who had ended his life in his early twenties without warning signs that any of us could read. Much like my own daughters, Callum had grown up on this campus, climbing its trees and finding candy in staff offices. The year before his death, I'd hired him to install hardwood floors in my kitchen, and since that time, I've thought about him whenever my bare feet touched the polished wood.

With two young and tired children at his funeral, I hadn't driven to the cemetery for the disposition of his ashes. But after the chapel service, I'd watched Callum's friends—a huddle of athletic twenty-somethings—jump onto their mountain bikes to carry his ashes over the mountain on the college trails to the cemetery.

Steve and I sat in silence for at least a full minute, and I recalled gathering in our front yard with my neighbors after we heard of Callum's death. On that cold January morning, my neighbors and I looked across Night Pasture toward his parents' house, uncertain how to assimilate the news into the life of this valley. We stood almost frozen together, staring at the red-brick house as if we could will away the truth.

"This place is for our community," Steve said. "It is the place where the people for whom buildings are named on campus are buried," he said. "You have to belong to the Warren Wilson community—the church or the college—to purchase a plot." Steve invited me to tour the cemetery with him and suggested I get in touch with former math professor Ray Stock, who kept a record of the cemetery plots, and Rodney, a retired staff member, who knew the history better than anyone.

"You'll want to make sure you have plenty of time to talk to Rodney," he said, smiling. We both knew Rodney's legendary reputation for radical hospitality, as well as the extended duration of his tales. Before I left Steve's office, we made plans to meet the following week for my first official tour of the burial ground.

The next week, I picked up Steve at the chapel, and we drove to tour the gravestones of the families who had built the college where I'd raised my children and taught hundreds of students. From the college archives and recent conversations with old-timers, I'd learned the school had a long history—beyond my time—of promoting experiential education without a wealth of financial resources, creating a community where students integrate on-campus work and community service with academic learning.

As Steve said, "This place has operated for a long time on a shoestring, a prayer, and a vision." It's one of only nine work colleges in the country—schools where students all work to maintain the campus. My daughter, Maya, was studying at a similar institution, Berea College in Kentucky, where she worked two jobs, catering events and tutoring Spanish.

"I want to go to a work college," she'd said when researching colleges, "just not one where every single person knows my mom."

On any given day, I taught environmental education in science classrooms in Witherspoon, hurried past the administrative buildings in Jensen, worked out in Devries gym, and got my computer fixed in the Bannerman IT building. With Steve, I walked past the headstones of people who led the college, marveling that I could have lived on campus for twenty years without seeing these same names on their headstones.

"I'm going to require each of my classes to find and visit the cemetery on their own," I said. "This is my new assignment related to sense of place."

We found the grave of Arthur Bannerman (1900–1976), the first college president after the school changed from a two-year college for boys from Appalachia to a four-year institution.

Next, we passed the resting place of Henry Jensen, who wrote the lyrics of the first alma mater. "These are the people who made the college possible," Steve said. "They built this place."

The one-acre cemetery was tucked into a northern nook of the campus, past the Verner Child Care Development Center and the Berea Baptist Church. From the cemetery, I could see the outline of the mountains and the pastures where college cows grazed. There were bluebird boxes on the fence line, similar to those perched in front of our rental duplex, bordering Night Pasture, which Annie Sky traversed each morning to catch the bus.

In my daily life on campus, I'd seen the signs on buildings without knowing these bodies had returned to the soil nearby. I felt like I'd been living in the midst of my own history without peripheral vision. We continued to walk past the graves: Sheriff Disu (March 30, 1982–Dec. 14, 2004).

"I remember when Sheriff died," I told Steve.

"Yes, he was a Muslim student whose siblings also went to Warren Wilson," Steve said. "An imam came from Winston-Salem and stood in the grave facing east. Sheriff was buried in a shroud, I'm told."

"Do these grave sites require vaults?" I asked, assuming vaults were underground, given the flat lay of the land and the lawn bordering the pasture.

"What do you mean?" he said.

"You know, like concrete liners in the grave," I said. "Many burial grounds require them so they can easily mow the grass, but some small cemeteries don't."

I'd first seen a vault during a recent visit to the conventional cemetery, where it sat graveside in preparation for a burial.

"Oh, we don't use vaults here," he said. "At least, I don't think so. I do so few full-body interments—I'm honestly not sure."

After walking on the even ground, I was fairly certain the graves in the campus cemetery had vaults.

Nearby, three older men were digging a hole for ashes that would be interred that Sunday. Standing by the grave was Ray Stock, the cemetery trustee I planned to interview later. He was part of a trio who managed the site, mowed the grass, and prepared for burials or interment of ashes. During the past week, I'd sent an email and left a voicemail asking to meet with him, but I hadn't heard back. As the other two men dug the hole, Steve called over to Ray to join us, and I chided Ray for screening my messages.

Short in stature but solid in presence, Ray Stock, eighty-five, was the epitome of his name. Compact and stocky, he looked less intimidating in his mid-eighties than I remembered him being when he was around campus. He held up his left arm in a brace and said he'd broken it in December. Ray had taught math and computer science at the college when my friend Lyn was in school; she'd thrown out her back while working in the campus garden and taken his entire course while lying on the classroom floor.

"I think you need to come to the Presbyterian church one Sunday and get you some religion," he said with a grin.

"Oh, I've been to the chapel many times, and I'm already taken by the Episcopalians," I bantered. "I go to All Souls Cathedral in town."

As the two other men shoveled, Ray looked over his shoulder: "Dead people are a hell of a lot of work," he said.

When I asked Ray about vaults, he said they do require them but use ones with the bottom opened so the container isn't sealed underground. Vaults seal the casket from the earth, while concrete burial liners can have three sides with the bottom open, so they maintain the flat ground for mowing but don't create a sealed barrier. While the terms are different, people often use the words interchangeably.

"It's harder to maintain the cemetery without the vault," Ray said. "The vault keeps the ground level so we can mow."

Steve raised his eyebrows, learning something new that day as well.

"But would the committee be open to discussing the option of not requiring a vault for one burial?" I asked. "When you bury a vault, you are burying almost 2,500 pounds of concrete for the sake of mowing. I've seen ways of closing the grave with a mound of dirt, so the soil doesn't settle below ground."

Ray didn't answer my question, so I let the issue drop. I'd found my dream cemetery less than one mile from where my children had grown up. The resting place was located on one of the most environmentally friendly campuses in the country. But I might not be able to be buried without a concrete bunker—something I was unwilling to do in my search for a sustainable end to my life.

Among the other things I learned as we spoke with Ray was that since 1957, the cemetery has been managed by volunteer trustees. But Ray added they were forming an incorporated entity called Warren Wilson Cemetery Inc. and aimed to transfer the land to this group. He continued to show me the burial grounds by the numbers.

"There are 240 plots, and 180 are spoken for," he said. "So there are 60 plots remaining."

"Well, that's good news, so there would be a plot if I wanted one," I said.

He nodded, continuing as if he were teaching math to students. "Each plot is four by ten feet, and the cost is $100 each," he noted. "Most people who buy a plot want a bigger headstone if they are being cremated. But with cremation, they could also get a small plaque with a memory stone in the ash garden."

The headstones by the graves were flat to the ground, one reason I hadn't noticed the small site when driving past. The cemetery seemed to blend into the landscape of the farm, tucked away next to the pasture.

"Do you have a digital copy of the plots?" I asked before he returned to his work. "I'm wondering where my neighbor Susan's site might be. She bought hers for $75 nearly twenty years ago."

As a response, Ray reminded me the price was now $100 for a grave site where you could bury two people's ashes or one full body. This seemed like a deal to me, and I wondered if I should purchase a plot for my children as well, just in case. The other day, Annie Sky had asked me, "What happens if I die and I don't have my wishes written down? Do the parents just get to decide what to do with my body?"

I'd never thought about the question, although it is relevant to anyone who has faced the death of a child. *At least,* I thought, *my daughter is starting to talk to me about death.* For the past month, I'd been carrying a miniature cardboard casket in the back of my car, a prop for presentations at an environmental education conference. We'd both forgotten about it until we had to move the casket to make room for her friends to sit in the backseat of the car. Her question was the first time she'd initiated dialogue about death, rather than turning up the radio as a way to shift the conversation to the living.

"I guess that's something I'll have to figure out this year," I'd told her. "But I think you're right; the parents would have to decide. Let me know if you have a preference."

While I pondered an impulse buy of additional plots for my daughters, Ray returned to his work. To continue the conversation, Steve suggested that I offer to visit Ray's house, where he could talk to me from the comfort of his recliner.

"He's getting older, and he's the one who keeps the records of the graves," he said. "He's given his life to this cemetery in so many ways."

Steve and I walked a few feet from the grave sites to the ash garden, where families could pour ashes over the large river rock, appropriately called "the big rock," which stood as tall as my waist. When Steve officiated a service, he would pour water from a glass pitcher after the family scattered the ashes over the big rock. The water mixed with ashes and then filtered into smaller river stones at the base of the memorial.

"If the family is Christian, I will share that the baptismal journey is complete in death as the water pours over the ashes," he said.

Nearby, smaller memory stones in a wall were engraved with the names of those whose ashes had been scattered. Some people I knew had already purchased their name plate in preparation for death.

Steve read aloud the names of those who'd been cremated: "Ah, Mildred—she worked in the campus bookstore and gave away books to students when they couldn't afford them. She would pay for the books herself but not tell anyone."

Before we left, the three cemetery trustees, including Ray, pushed the wheelbarrow filled with dirt up to their trucks.

"Now don't go writing that it took three old men to push a wheelbarrow!" one yelled to me.

I laughed out loud as Steve continued to narrate the college history for me, one grave marker at a time. We passed the couple for whom our gym was named: Samuel DeVries (1911–1991) and Evelyn DeVries (1910–2012). "They lived on campus on Daisy Hill, and when she got to be 100, she'd tell me, 'I'm older than my weight!'" said Steve.

As we walked back to the car, I looked from the grave sites to the skyline, where the Appalachian Mountains, lush with summertime green, surrounded this small parcel of land over the long course of time.

The next week, I drove past the ranch-style 1970s homes on College View in search of Ray Stock's house on campus. I'd had time to compare the low cost of plots at the Warren Wilson Cemetery. The average cost of a burial plot in the United States recently ranged from $200 to $2,000 in public cemeteries and from $2,000 to $5,000 in private cemeteries. Burial grounds charged additional interment fees to open and close the grave: $350 to $1,000 in public cemeteries and $600 to $3,000 in private ones.

When Ray opened his front door, he led me to the dining room table, where he'd laid out a piece of yellow poster board with a grid that depicted the locations of each person whose body or cremated remains were buried in the cemetery. On the poster, I could see names written in light pencil, including the heavyweights of the college's history: Billy Ed Wheeler, Henry Jensen, Thekla Jensen, Arthur Bannerman, Lucile Bannerman.

I sat down in a wooden chair in front of the poster and asked if this was the only copy of the grid.

"This is the only hard copy of the plots," Ray said. "I have an Excel spreadsheet with the names as well."

"Does anyone else have access to the spreadsheet?" I asked.

"I don't know if anyone else has it," said Ray. Then, turning to his agenda, he began to explain the eligibility requirements for buying plots in the cemetery.

"Faculty and staff and trustees and their immediate family members are eligible to buy plots for $100 or to spread their ashes and buy a plaque and a memory stone for $50," he said.

"So this would include me and my children, right?" I asked. He nodded and continued.

"Then members of the Warren Wilson Presbyterian Church and their immediate family are also eligible to buy plots. Alumni can purchase a memory stone in the ash garden for $50 or spread their ashes at no cost," he said, adding that anyone eligible could spread their ashes without having to pay. Ray estimated that about forty people's ashes had been dispersed, although if they didn't purchase a stone, there was no record.

"I don't know why someone would want to spread ashes and not buy a memory stone," he said, shaking his head. "No one will have any record of the death if there isn't a name."

We then talked about reasons why someone might decide against buying a stone, such as when the pain of a child's untimely death might be hard to memorialize forever.

Standing up to get a better view of the poster, I put on my reading glasses and tried to find the plots owned by my neighbor. With a quick scan, I located her plots at H18 and H17, to the left of Hugo Shorpe, who was already buried.

"I'm going to take Susan to the cemetery to see the location of her plots," I said. "While she could technically sell hers to me, I'd rather buy my own than have her ex-husband's plot for $25 less, since she bought hers."

Ray nodded in complete agreement. We'd both been divorced and didn't want that kind of karma going with us to the grave.

On the gridlines, I saw the name "Sheriff Disu," the Muslim student who had died ten years earlier.

"Could you tell me about Sheriff's burial?" I asked. I knew Muslims wouldn't bury with a vault.

"We made an exception in his case, especially since they needed to bury the body quickly according to Muslim tradition, within twenty-four hours," he said. "His body was shrouded, and the trustees dug the hole and waived the cost, since the family didn't have financial resources. The imam positioned his body so he was facing east."

So there was at least one burial in this cemetery without a vault.

"We do all the landscaping, dig the holes for the ashes, and care for the cemetery, since we don't make enough money from the sale of plots to hire someone to help," Ray said. "I had a riding lawnmower, and the work had to get done. So I started to care for the cemetery."

And there you have it: Ray had spent thirty years caretaking a sacred space because he owned a mower and the work needed doing, the epitome of the Warren Wilson spirit.

"Lyn and I talked about buying plots next to each other and putting a phone line in between our headstones," I joked.

I decided to bring up the contentious issue of the vaults again—the one issue getting in the way of my own natural burial on college land—because I wanted to minimize the carbon footprint of my death. I told Ray about my own father's natural burial in our neighborhood cemetery and how his friend had built a pine casket for free. Ray said he'd never seen a casket sold for less than $12,000.

Ray was a natural-burial Eeyore to my optimistic Christopher Robin. I'd loved the green burial grounds at Carolina Memorial Sanctuary, with its mission of conservation and community, but this cemetery at Warren Wilson felt like a touchstone to me.

As a next step with Ray, I described conventional cemeteries that allow vaults and embalming, like Washington's Congressional Cemetery, but *also* include selected plots for natural burials.

"You can have vaults for most of the cemetery but have a section for natural burial," I said. "It's called a hybrid burial ground. Besides, most people with plots at the Warren Wilson Cemetery are getting cremated at this point anyway."

I was obviously trying to make a case that green burial might be an exception to growing numbers of people choosing cremation. But Ray looked unconvinced.

"We know sea levels are rising; it's the hottest July on record; hurricanes are wreaking havoc on the coasts; the Amazon is burning," I said, my voice racing faster, my frantic parental voice kicking in. "Natural burial is one small action I can take. We can actually decrease the carbon emissions of our own deaths and regenerate the soil on campus. It's not like boycotting Exxon Mobil, but it's a start in the right direction."

Even by my own accounting, I was feeling desperate, almost trapped. There were three men, two of whom I'd barely met, possibly standing in the way of my hopes for my own burial. Certainly, I had the choice of natural burial at Carolina Memorial Sanctuary, but regardless of my resting place, I also wanted the cemetery on campus to reflect our shared values of sustainability.

"Would the cemetery trustees be open to a conversation about the environmental benefits of not using vaults?" I asked. "I'm not asking you to make any decisions, but we could talk about methods of burial so the maintenance isn't a problem. There's a Warren Wilson graduate in charge of the conservation cemetery here in Mills River, the only one in North Carolina. She could even come and talk with us."

"We won't take up the decision of a vault or no vault right now," Ray said, closing the discussion. Although he was five foot two, he projected the impression of a six-foot-tall man by virtue of his confidence and his grounding.

"I'm not asking for a decision," I repeated. "I just want to have a dialogue." I could see Ray's shoulders lowering into his chair.

Then I sank to a new low as a southern-born climate feminist: "Look, we can have a talk, and I'll cook y'all dinner, and we can all have a beer or a glass of wine," I said. He'd told me earlier that he was waiting for a shipment of wine from his children, who were coming to visit for his birthday.

At that moment, thunder boomed outside, and a downpour pelted the windows. I packed up my bag and asked if he would show me a picture of his children. Ray picked up a Shutterfly book and shared each photo of his past birthdays with all his grandchildren, four children, and their spouses. His special friend, retired from the admissions office, was in the picture, seated next to him. We talked about each child and grandchild, and he told me where they were going to school and what they were doing for work. In the end, I knew this connection mattered more than my green-burial campaign, at least in this moment.

Although I would continue to talk to Ray about why a green burial made sense at our college, I was determined to start bringing my classes to the cemetery, showing my students history in one place. This story would have another chapter, and as with my life, I wasn't sure how it would end. But I would tell my students about this man who'd volunteered hours of his life to honor the graves of those who built the college. In those days, I couldn't have predicted I'd continue a campaign for natural burial so close to home during a pandemic that would touch the entire world.

Chapter 7

THE CONTAINER STORE

Shrouds, Pine Caskets, Mushroom Suits, and Cardboard Boxes

THE WATER SLOSHED AROUND our feet. But the band continued playing on the stage of Pisgah Brewery, nestled in a nondescript industrial building behind the Bethel Baptist Church in Black Mountain, North Carolina. The scene felt like a remix of Noah's ark and a Brews Cruise. Bartenders used sponge mops to soak up the water seeping onto the floor during the downpour, yet the invading rain didn't stop the taps or silence the band, led by my college roommate and songwriter Liz Teague. When she wasn't playing original tunes like "Do What Your Momma Says," she worked as a city planner for a nearby mountain town facing the dual challenges of development and dramatic weather events—like flooding exacerbated by climate change.

Liz and I had learned to play the guitar together in college, and while I rarely played after graduation, she now recorded albums and performed gigs around town. My minor—and somewhat comical—contribution to the band involved leading the crowd in hand motions pantomiming the lyrics of a song called "You Carry My Love." Onstage, I'd point to the crowd to signal "you," hold my hands in the air for "carry," and cross my arms over my heart to portray "love," all while singing the words at the top of my lungs (but without a mic). Now I was looking for the best way for my daughters to carry my love after I was gone.

Even with the half inch of water on the concrete floor, nobody made a move to stop the music or head back into the raging storm.

"Don't you think they should avoid those electrical outlets?" I asked Steve, Liz's husband, who was standing nearby on the taproom's damp floor.

In a deadpan tone, he replied, "They need a wet vac," which reminded me this wasn't my problem to solve.

Fifteen years before, Liz and Steve had made the trek from North Carolina to Alabama to attend my dad's funeral. Together, we sang "All Things Bright and Beautiful" in the Episcopal church with my father's pine casket in front of the altar.

"If Jeff can't build a casket, you can always use a cardboard box," my father had written in his two-page directive. "A refrigerator box from Mills Appliance Store would work just fine."

After seeing the environmental benefits of green burial over cremation, I faced the decision often made by families in a funeral home soon after death: What did I want my family to bury my body in? A pine casket, a linen shroud, a wicker basket, a cardboard coffin? At the time, I didn't realize my friends with the band would help me with that decision, honoring my parents and the earth at the same time.

* * *

Before driving to the brewery that night, I'd attended two funerals, one in the morning at my church for a ninety-four-year-old doctor named Oz, and a second in the college chapel for a choir director and organist who had died from cancer in his sixties. At the Cathedral of All Souls, the Rt. Rev. Brian Cole—or "Bishop Brian," as he was known—preached the eulogy and reminded us, "Today our task is to proclaim resurrection, to declare that life has swallowed up death. We say Oz's life has changed, not ended." A wooden box about the size of a breadbox sat on a stand in front of the altar, holding a plastic bag with the cremated remains.

After the service, we walked into the small courtyard of the church, where Brian interred the ashes, pouring the pulverized bones into a shallow hole in the ground about the depth of my hand. There were no markers for the dead. I'd attended this church for ten years before realizing that hundreds of remains were buried in this small flower bed graced only by a sculpture of Saint Francis. In the drizzling rain, I stood next to Oz's family and friends, who used their bare palms to place soil on top of the ashes. After the last person came forward, the clouds lifted, and sun rays filtered into the courtyard. Without words, the group of about twenty people looked to the sky in what felt like a collective prayer.

What happens when we gather together—as one body—to mourn someone whose own body rests in an urn, a shroud, a coffin before us? I think the idea in the Episcopal liturgy that "life is changed, not ended" has something to do with the lyrics of Liz's song. Somehow, we must carry the love from the past as we create meaning from the absence and the end. For me, learning *how* was the point of it all.

"What do you think about helping me build my own casket?" I asked Steve as the band in the brewery took a short break.

Before he could respond, I answered my own question out loud: "It doesn't make sense to do that now, as I have nowhere to store it."

"I'd be glad to help, but I don't know where you'd put a big ol' wooden box in your small house," he agreed.

Besides the lack of storage, the other kink in the plan was that I wasn't exactly a DIY kind of gal, even though both my mother and father could sew, build, fix, and craft. I wanted to be as handy as my students, who could spin yarn from sheep wool and concoct salves from fresh herbs, but I tended to find other people to fix things when they were broken. Living on a college campus with undergrads who offered labor for affordable prices was another enabling factor.

"Every town once had a furniture maker who could build a coffin in a day or so," Steve said. "But we've lost that practice. Now you have to buy an expensive one from a funeral home or get cremated. Or you can build it yourself."

"Isn't it crazy that back then, coffins had six or eight sides to more closely resemble the human form?" I added. "Later coffins changed to have only four sides—more like a rectangular box—hiding the fact there was a body inside!"

Steve nodded, amused by my enthusiasm about the details of disposition.

Before this research, I'd assumed coffins and caskets referred to the same burial container, but I'd learned they differed according to the number of sides. In the 1700s, coffins—with six sides tapered at the head and foot—became popular. Before then, most people were buried in a shroud or a sheet. The coffin replaced the shroud because people believed that the shell could "protect" the body for resurrection on Judgment Day—an idea that came from the Christian tradition. Ironically, Jesus was laid to rest in only a shroud. The Buddha was cremated.

Today, the tapered hexagon of the coffin is less commonly used, after caskets with four sides became more popular during the American Civil War. By the end of the 1800s, a simple pine box wasn't sufficient for those with resources, and manufacturers designed the casket, a word that originally referred to a jewelry box, from the Old French word *cassette*. With the rise in metal cookstoves and the middle class, the metal burial container began to be marketed as providing even greater protection from nature.

From my research, I knew that no federal or state laws required purchasing a casket from a funeral director. You could buy one from Costco, make one yourself, or order it online. A cremation container—a cardboard box used to cremate a body—or a biodegradable cardboard coffin also could work just fine. And as my father suggested, a refrigerator box was legal too.

"A four-sided box would be easier to build," I told Steve after the next song ended. "Maybe you could just build my casket for me, like my dad's friend did?" I asked. "The act was super meaningful for him after my father died."

This solution was a bit of a cop-out on my part. Steve had a modest wood shop, where he'd refurbished my childhood toboggan so my daughters and I could sled down the hills during the now-rare snowy days. In my proposal to him, I'd seen a way to avoid constructing and lugging a casket with me for the rest of my life.

"Sure, I could build your casket after your death," he said, "but I'm a fifty-six-year-old male who might die before you."

"Well, you never know," I said. "I need to decide on something to hold my body, and Annie Sky doesn't want me to have only a shroud. She'd like some kind of container."

"You're still planning your funeral based on the opinion of a fourteen-year-old?" he said.

"It appears that way, doesn't it?" I said.

"Hey, it's your funeral," he said. "But maybe you could start by building something simpler, like a bluebird box where birds could nest in your yard."

On that note, Liz's band began to play. My college roommate still had a smile that wouldn't end and dimples that made me feel more hopeful about our world. The bartenders returned to serving brews; the floor was mopped up, the water dispersed. Once again, life was changed, not ended.

The same week, I fell in love with a shroud, the way I might have swooned at a wedding dress (if I ever planned on getting married again, which, for the record, I didn't). It was made of simple blue-gray linen by a local fashion-designer-turned-antique-dealer named Christi Whitely, owner of Shrouded House. Staring at my laptop screen, I gushed over the frayed edges of the linen, the natural dye, the fabric ties that would fasten around the body. I wanted to wear it while I was living, as well as when I died.

The designer had collaborated with Carol Motley, co-owner of a company called Mourning Dove Studios, which sells products like sturdy cardboard caskets and shrouds for natural burial. Carol had seen a need for a piece of fabric to wrap around a body that could accommodate a range of body types—a one-size-fits-all pattern—and her friend Christi was up to the task. In fact, while she experimented with designs, her husband "tried on" the shrouds, so she could get a sense of the fit. (Christi made him smile in each of the photos.)

When folded, the shrouds looked like they belonged in an Eileen Fisher catalog, given the calming hue of the linen and the soft edges of the fabric, and I wanted to be surrounded by that material and placed into the ground. Shopping online, I felt like a participant in the reality show *Say Yes to the Dress*, although I didn't plan to try this on for size. In this pattern, a specific stitch in the fabric could be shortened or lengthened to work with different body types, like stretchable waistlines. The simple aesthetic reminded me of my mother's sewing closet, where she'd kept stacks of cotton and designed embroidered dresses for Maya as a toddler. If my mother had had more time to plan the ending to her life, I could imagine her willingness to sew her own shroud.

Shroud burial is an ancient tradition that was practiced by some early Romans and Hebrews and has been continued by Orthodox Israeli Jews and some Muslims. Now shrouds have made a comeback with natural and conservation burial. To comply with Green Burial Council standards, a shroud must be made from biodegradable and nontoxic material. A standard bedsheet would work, assuming it was made of cotton, linen, silk, or wool and would decompose. The Shrouded House fabric included ties and handles to facilitate carrying the body, which struck me as being practical. Books like *The Green Burial Guidebook* include instructions for sewing a shroud, though it seemed unlikely I would ever want to do it myself, as our sewing machine had been in the attic for years. As I'd learned at Carolina Memorial Sanctuary, a common practice is to tuck fresh flowers or handwritten notes into the ties of the shroud before burial.

"Come and look at this amazing shroud!" I called to Annie Sky after dinner. Trying to appeal to her sense of fashion, I noted the quality of the fabric and its shabby chic aesthetic. "Doesn't it look like me?"

We were both sipping hot ginger tea in the kitchen of our duplex on campus with a view of the Swannanoa Valley.

"It's okay," she said, glancing at my laptop.

"Well, it's not cheap," I said. "The price is $300, but if I bought it for a birthday present to myself, you wouldn't have to pay anything to cover my body when I die." She babysat each month to pay her $40 phone bill and focused on saving money for future expenses.

"We wrapped my dad in my mother's linen tablecloths, which were free, but then we put his body in the pine casket," I said. "But with this shroud, you wouldn't even need a casket." At the conservation cemetery, I'd volunteered at a burial that only used a shroud, although the body was carried in a wicker open basket. At the grave site, about five people lifted the shrouded body and placed it into the grave, and the staff could reuse the basket for the next shroud burial. Families or individuals could buy this specific shroud from the staff at Carolina Memorial or from Mourning Dove Studio online.

"I want a casket for you and not just a shroud," she said. "And why can't we get a nice one, like from a funeral home?"

"A fancy casket would cost thousands of dollars," I said, "And that money could go toward buying that car you want." Without shame, I was appealing to materialistic priorities to negotiate with a fourteen-year-old about my own funeral. I didn't mention that Carolina Memorial Sanctuary wouldn't accept a casket with metal and exotic wood, since those materials wouldn't degrade.

"Okay, let's go with the pine casket," she said.

"Maybe I could get the shroud and keep it, because you might not want the casket for me once you're older," I said. "But if you do, there are places where you can buy one. I just don't know if I want to build one and keep it in the basement for years."

"Yeah, and it would get totally moldy in the basement," she said, plugging in her earbuds and leaving me to stare at the crisp linen shroud on the screen.

The day after two funerals and a band gig, Steve sent me a Facebook message: "Interesting stuff," he wrote above a link with a photo of a coffin. "If you make a simple design, you could use it as a shelf/bookcase until you need it." I noted his use of the pronoun "you," which signified I would be doing the building with his guidance.

The YouTube video he'd attached, entitled "How to Build a Coffin," was created by the owner of Nature's Caskets, a business specializing in hexagonal coffins. I'd come across a similar company in North Carolina called Piedmont Pine Caskets, but the shipping alone would be expensive. And if I needed to be buried within several days to avoid embalming, my daughters would need a casket or a coffin within a day or two.

I was headed to class but replied to Steve's message: "I have a cheap wooden bookshelf that I could give to students," I said. "It would make sense if I could actually use the casket. Maybe we should build my casket, I mean my bookshelf!"

He texted back the word "Ha!" followed by his signature thumbs-up emoji.

But did I really want to build my own casket when I was a healthy fifty-something with a full life, a tiny house, and no plan for where I could afford to live when I was too old to teach at the college?

That night on Instagram, Lyn sent me the link to a BBC video about an informal organization called the Coffin Club, a group of senior citizens who would gather every Wednesday afternoon to decorate their own caskets. Started by a former palliative-care nurse, the club had the purpose of personalizing burial containers but also providing social activities for elders.

"It's expensive to buy a casket from a funeral home," said the founder, "so we buy direct from a manufacturer. Everyone can afford it, and they can decorate the casket."

One woman with deep wrinkles on her face and a broad smile showed off her casket decoupaged with pictures of cats and kittens. A man who'd been a race car driver decorated his end-of-life box with checkered black-and-white racing flags.

"I've gotten calls from all over the world," said the founder. "It's just taken off, and people have started their own coffin clubs for the same purpose."

I wasn't clear from the video where folks were storing these caskets, which weren't biodegradable, but the story highlighted another trend I'd discovered in the natural-burial movement: personalizing cardboard coffins. Carolina Memorial Sanctuary in Asheville sold hexagonal cardboard caskets for $300, designed by Mourning Dove Studio. These cardboard caskets were strong, easy to transport, and made from bleach-free cardboard. Families often used markers, paints, and pictures to memorialize and share messages about the person who had died. During a home funeral, decorating the burial container could involve young and old in this ritual.

"I don't think I want an arts-and-crafts project at my home funeral," I texted Lyn. "I mean, I like it for other people, but I just want a simple container—no colorful rainbows or mountain scenes. The plain cardboard coffins are actually quite beautiful to me."

"Got it," she said.

"And with this Coffin Club . . . where are they storing the caskets?" I wrote.

"They aren't coffin up that information," she punned—no emoji needed.

Biodegradable burial containers are more affordable, simpler, and better for the environment than many of their counterparts I'd seen in funeral homes. Conventional burial containers, from caskets to urns, have a higher carbon footprint than biodegradable ones and can often leach toxic chemicals—such as from stains or varnishes—into the soil and water.

For my budget, it mattered that the average cost of a green burial container was $275 to $3,000, compared with a range of $2,000 to $20,000 for a conventional casket. The average cost of a casket of metal or imported wood was $2,000 to $5,000, and the use of tropical and exotic woods often contributes to deforestation. In contrast, many of the caskets used in natural burial were made from fast-growing pine.

Of course, my father's casket, built from scrap pine salvaged by a friend, had been free, including the handles made from sailing ties on each side of the box. The Carolina Memorial Sanctuary website includes a three-part blog series with links to resources for pine caskets (none including a bookshelf), wicker baskets, and biodegradable urns. Their advice is to call the natural burial ground closest to you, as staff can recommend a funeral home or a local woodworker selling green caskets, approved by the Green Burial Council. Many of these woodworkers use a small number of screws and nails, while others, such as Aldergrove Caskets in North Carolina, advertise a "plain pine box" crafted with joinery as the technique to fasten the sides without screws or nails.

When I saw a video of owner John Jull building these caskets, the product looked just like the prototype my father had built when I was a child. I emailed for pricing, and the simple but beautiful box cost $1,800 with $230 in shipping costs. This was too much money for me, despite the quality craftmanship of the product. The owner of the company was part of the Carolina Casket Collective, a regional group of casket makers.

I wrote back to see if they designed any less expensive models from pine. His wife replied within the hour: "I've spoken with John, and he's taken the attached photos of the simplified casket he is currently at work on. As you see pictured, it would have no handles; it would need to be carried using ropes or straps."

The photos looked exactly like the casket my father's friend had built for him: a sturdy, smooth, and simple box intended for the end of life.

"John will be finishing up the one in the shop in the next couple of days," she responded. "Let us know if you have any other questions at all. I am so sorry you are in the market for a casket. It is such a trying time."

This work can be relational, I thought, when it's done with care and communication. The cost for this simplified casket would be $1,350 plus $230 for shipping. Having handles would be critical, but this couple also seemed like they would work with and communicate with their customers.

I found a cheaper pine casket for $600 from a company in New Mexico called the Old Pine Box, but shipping a casket across the country was cost-prohibitive. Other sites, like the Natural Burial Company, sell DIY casket kits. I also found free online plans for building your own simple pine box for burial.

After this research, my perfect shroud felt even more practical, with the backup of a pine casket built with a friend when I was closer to death or a $300 cardboard casket available through the conservation cemetery or Mourning Dove Studios.

The latest post I found on Carolina Memorial Sanctuary's blog about biodegradable burial containers highlighted the "Strange and Unusual." These included human composting (which I explore in chapter 9), biodegradable egg-shaped burial pods that encase the body in a fetal position and are buried in the soil, and the infinity mushroom burial suit. While the burial pod wasn't ready for market yet in this country, I was able to check out the mushroom burial shroud, the same product that actor Luke Perry was buried in at age fifty-two.

A friend introduced me to artist and printmaker Maria (pronounced Moriah) Epes, who lives in a small, quaint 1895 farmhouse in Asheville with a simple aesthetic of bright white and deep blue inside the home and a lush green garden outside. I'd planned to stay for an hour and ended up visiting for three, learning how art can help to prepare us for death, especially with a shroud burial. Maria's father died when she was only eleven years old, an early experience that affected her artistic fascination with death.

"In my sixties, I knew I needed to make my end-of-life decisions, and it was important for me to have an art project to get myself to do it," she said. So she consulted a lawyer to help her write a will, health care power of attorney, and living will, while integrating art into the process.

In her living room that served as a studio, she'd spread out her Coeio infinity shroud, which looked similar to the crisp muslin curtains adorning the windows. The shroud was white cotton, infused with mushroom spores, which promote decomposition. Maria kept it in a green compostable bag in a dry place, as per the directions from Jae Rhim Lee, the founder and CEO of Coeio. She'd made me a copy of the instructions:

"Thank you so much for supporting our shared mission to celebrate life by making death an ecologically sound and beneficial moment!" The letter gives directions for wrapping the body in the shroud, which include corresponding numbers printed on the fabric, as well as straps for carrying the body.

Maria showed me an original print that she wanted each person at her funeral to receive. She pointed to her signature on the artwork, ME•XX—her initials, followed by XX for woman, as she saw the interplay between feminism, birth, and death.

"It's almost like a party favor," I said, marveling that she was the first vibrant, healthy person I'd met investing so much time in preparing for her death.

"These will be buried with me," she said, pointing to three clay amulets, figures made with pigment from Jones Mountain on the Warren Wilson College campus, where I lived. Then she showed me another shroud sewn in rectangles from her grandmother's linens.

"I have similar linen place mats from my mother and grandmother," I said.

"I asked each of my friends to give me one phrase that uses the word *dead*, and then I painted those words on the linens and sewed them together," she said. "Look at what they came up with!" She read aloud from the white material: deadpan, to die for, deadhead, dead reckoning.

This artist had created a map for her body—with three shrouds to surround her at death.

The first shroud would be a blue sienna one, made from prints depicting X-rays of her skeletal system. The pieces of blue fabric, adorned with white images of her bones, hung from a clothesline in the studio. This would be the first layer, followed by the white linen one with words from friends, and finally the mushroom-infused shroud.

"All of us are full of toxins, and I did a lot of research and saved up the $1,500 to buy this shroud so my body can truly become compost."

"Couldn't you just pour some mushroom spores on your other shrouds and get the same effect?" I asked, repeating a skeptical comment I'd heard from Cassie, the manager at Carolina Memorial Sanctuary. (She'd called the infinity shroud "BS" during one of our meetings, due to its cost and what she described as undocumented claims.)

Maria explained that Jae Rhim Lee had proprietary rights for the combination of mushroom spores and bacteria that help the body decompose and remove toxins. I reached out to touch the soft inside lining of the mushroom shroud, which felt like fleece to me.

As if reading my mind, she said, "You know, Cassie is not a fan of this infinity shroud. She doesn't think it's worth the price. But I took my time doing the research and wanted to make the investment."

Maria intended to purchase a plot at the conservation sanctuary and spread out the payments over two years.

"After all this research, I've decided that natural burial in an infinity body suit is the way I'll go," she said. "Alkaline hydrolysis seems like it leaves the least impact, but there isn't a facility in our area yet, although it's legal in North Carolina. It uses old-fashioned lye, and the body decomposes to liquid. So I'm now committed to the conservation cemetery."

As we talked, a black Oriental cat named Vladimir rubbed against my ankle, unaware of my allergies. This peaceful home adorned with her artworks also included catwalks and play structures for her kittens.

Lastly, Maria had created a portfolio called "Time's Up," containing all the information her niece would need to close out her accounts, from passwords to logistics about her house and insurance. In a creation entitled "Time's End," a black clamshell case she'd made held instructions for what to do with her body, the details about her funeral, and how to get the death certificate.

Before I spent the afternoon with Maria, it hadn't occurred to me that the end of our everyday lives could be a work of art.

"Why don't you just build the prototype of your casket, like your dad?" asked Annie, my running partner, who taught education courses at the college.

"You'd have the model for Annie Sky and Maya," she said. "They'd know what you liked, and you could keep jewelry inside the small box."

During the hour of our run, we discussed work, relationships, and children, but we always lingered in my gravel driveway to tackle any emerging final topics. We'd been running together on Saturday mornings for more than fifteen years. With her, I felt known and at home.

"A prototype, hmmm?" I said. "I guess the point would be to experience building a casket on a smaller scale."

"You know Steve can build a small casket, since he built the one for the goldfish before we cremated it," Annie said, grinning about the funeral held on Liz and Steve's back porch for a beloved beta fish. "What was the name of that fish?"

"Diane," I said. "How could you forget? Well, you didn't know her as long as we did."

Indeed, Steve had built an intricate engraved wooden casket shaped like a fish to house Diane's body after an unusually long life of five years, including a period of minor neglect by her first owner, their college-age daughter. We'd had a potluck after cremating her casket and body in the outdoor fire.

"Maybe in years ahead, Maya and Annie Sky might end up feeling comfortable with just a shroud—or a shroud and the cardboard coffin," I said. "If they want a wooden casket, there are options for a simple pine box. I'm not sure I need a model to send that message, although I did love Diane's little casket!"

The end of our trail run often brought some sense of resolution for me, the lifting of unease, even when it didn't last for long. Much like facing the climate crisis in solidarity with my students, preparing for death with my friends and family created a structure of support that didn't require a saw, a hammer, or a piece of pine. These conversations might not change the world in a heartbeat, but they could create a prototype of the connections we'd need down the road. And if I could give my daughters a pathway to carry my love, perhaps they could craft meaning from the changes to come.

Chapter 8

ASHES TO ASHES, DUST TO DUST

From Flame Cremation to Aquamation

In my midtwenties, I smelled the metallic scent of scorched flesh and saw charred limbs extending from a funeral pyre by a river. Men in white robes stoked the burning straw and wood while family members stood near their beloved, who was wrapped in a shroud, shape shifting from orange flames into gray ashes.

After teaching in the Peace Corps, I'd traveled from the Central African Republic to Nepal, where I witnessed human bodies burning, the smoke a backdrop to the bustle of everyday life. In my weathered Lonely Planet guidebook, I'd read that Hindus believe cremation releases the body into the five elements: air, water, fire, earth, and sky. When the body perishes, the soul persists. I couldn't claim to understand the religious ritual by the Bagmati River, which flows into the sacred Ganges. But this public cremation reflected an integration of life and death I wouldn't experience until I lifted my father's body into his home-crafted pine casket.

Decades later, I learned about the town of Crestone, Colorado, home to the only community open-air funeral pyre in the United States. Residents and landowners in the county can use the pyre for their funerals at a cost of about $500 to cover one-third of a cord of wood and the fire department's time, among other minor expenses. The deceased are covered in a shroud, carried on a stretcher, and placed on steel grates set into a concrete hearth lined with bricks. Surrounded by juniper logs and branches, the body burns for four to five hours in the presence of the Sangre de Cristo Mountains. The volunteer-run Crestone End of Life Project oversees the cremations and also facilitates trainings on death and dying much like the Center for End of Life Transitions in my home of Asheville. When a body turns to ash in open air, there is no denying death.

Most people in the United States don't have the option of an open-air cremation, but more than half choose cremation, and that number is expected to rise. During flame cremation, bodies are burned inside furnaces in crematories and funeral homes. This process typically doesn't involve the family, and it uses fossil fuels that contribute to the warming of our planet. The cremation of one body produces 250 to 350 pounds of carbon dioxide, the equivalent of a thousand-mile car trip. Across the world, we often handle our bodies after death in ways that harm the earth—sometimes because of religious and cultural beliefs and other times because there aren't viable alternatives.

After flame cremation, the remains of the body—also called cremains—are gritty and gray, with a texture similar to sand. In a cremation chamber or furnace, a cardboard box holding the body is the first to burn. Our bodies are 60 percent water, which then evaporates quickly. The soft tissue burns for about two hours, until what's left are bone fragments. These bones are raked and swept up from the furnace and then ground into the course substance we call ashes. Crematories and funeral homes place the ashes in a plastic bag, often put into an urn before being returned to the next of kin.

The final wishes I'd written in 2010, five years after my father's natural burial, directed my daughters to cremate my body. With this decision, I'd prioritized the convenience and low cost of basic cremation.

In the meantime, I'd learned more about its environmental impacts, the primary reason for reevaluating my directives. Cremations in the United States produce 250,000 tons of carbon dioxide emissions each year, the equivalent of burning 30 million gallons of gas. Most of my friends wanted to be cremated and assumed this choice was more sustainable than conventional burial with a concrete vault and embalming, which is true. But it also isn't a green practice. I was ready to explore alternatives to flame cremation that use smaller amounts of fossil fuels and also engage the family in after-death care.

"How is one person's choice to cremate their body going to make a difference with the climate?" asked the electric guitarist for Liz's band. Both in our fifties, we were once again hanging out at a local brewery during a break in the set, headlined by the Liz Teague Band.

It was a good question, and I'd certainly never talked about death and dying with a gray-haired guitarist who held a day job as an engineer.

"One person acting alone might not make a huge difference, but I'm betting on the momentum created when people act together," I said. "Think about the demand for midwives and natural childbirth by our generation as compared to our parents."

The musician didn't look convinced, but this was an age-old discussion within circles of climate activists. Should we attack root causes of the climate emergency and forget about individual behavior changes? It turns out the small daily choices we make can have a collective impact on the climate, even if scientists say individual actions can get us only 30 or 40 percent of the way to safety. The rest falls on governments and companies and systemic changes, often with public pressure, away from our reliance on fossil fuels. Only 100 companies in the world are responsible for 71 percent of global greenhouse-gas emissions, and the US military is the institution that consumes the most fossil fuels in the world.

As a leading climate scientist and a Christian, Dr. Katharine Hayhoe claims that one of the most important things we can do about climate change is to talk about it and connect the dots between the climate and what people already care about. In 2019, she said three-quarters of people in the United States heard someone talk about climate only once or twice a year. As natural disasters affect more and more people, this frequency will change. Like her, I believe in pressuring institutional systems and also taking individual action at home.

During this year of research, I examined the impacts of cremation to ensure I was making an informed choice, even though cremation would likely end up as my plan B if I couldn't have a natural burial or my death happened far from home. Planning for a green funeral and disposition matters to me, in addition to voting politicians into power to confront the climate crisis and supporting my students pushing government at all levels toward climate action.

The percentage of people choosing cremation in the United States is expected to increase from the current rate of 50 percent to 80 percent by 2040. This is a remarkable shift in US cultural practices, given that only 10 percent of the deceased were cremated in the 1980s. It's also an example of how change happens with momentum over time.

One-third of all funeral homes now operate cremation equipment. A significant reason for this growth in the industry is the relatively low cost to consumers. When comparison shopping, I learned that direct cremation—covered by the $995 price at the low-cost crematory near my church—includes transport of the body, filing the paperwork and death certificate, the cremation, and return of ashes to the family. That price didn't cover a memorial service, for example.

In addition, cremation doesn't require advance planning, and the remains are portable, which is appealing when people frequently move. In many other countries, cremation remains a popular choice as well: 99.9 percent of people in Japan choose it, as do 85 percent in Switzerland and 80 percent in Thailand.

When I met with funeral directors, I'd continued to inquire—in a polite Southern way—about watching a cremation for research purposes only. "We still can't let you observe a cremation, for privacy reasons," said Stanley Combs, funeral director at Asheville Mortuary Services and Asheville Area Alternatives. I'd returned to this budget funeral home, where cremation without a funeral service costs less than $1,000.

Before our meeting, I'd waited in a small room where open shelving stocked a legacy silver urn ($195), a green-and-yellow John Deere urn ($165), and a pillow and blanket adorned with the photo of a striking silver-haired man in a cowboy hat, who was identified as Robert Gardner (1957–2016).

After talking to Stanley, I'd learned that the cost for cremation in our region ranges from a low of $895 in a nearby town to $4,000 at an upscale funeral home with an expansive cremation package, which put the entire range below the $10,000 average price tag for a conventional burial. The Funeral Consumers Alliance says a reasonable rate for direct cremation without additional services is $800 to $1,200. The facility at Asheville Area Alternative cremates about 150 bodies a year for other funeral homes that don't have the equipment, in addition to 400 direct cremations. (Imagine the scenario of a mortuary paying a low-cost facility to cremate a body and then inflating the price for the consumer, which does happen.)

"What do you think about the environmental impact of cremation?" I'd asked Stanley.

"Cremation puts mercury and carbon dioxide into the air and uses fossil fuels," he said without any hesitation. "Sure, it's better than conventional burial that puts formaldehyde from embalming and a concrete vault into the ground. But people think it's good for the earth, and it's actually not."

When we'd talked earlier about natural burial, Stanley had seemed just as honest as during our discussions about cremation, which his company promotes as an affordable choice.

"Conservation cemeteries seem like the best option that's environmentally friendly," he continued. "This would be the number-one choice in terms of decreasing impact on the earth. You don't even need a casket if you just use a shroud. So you don't have the transport and manufacturing costs of an expensive casket, and you save fossil fuels that way too."

"So what are your arrangements for your own body?" I asked.

"Given the business, my arrangements are for cremation with a service at our funeral home," he said. "If my son wants to throw my ashes in the Swannanoa River, he can. If he wants to put me in an urn, he can do that. I want him to decide that part."

This river meandered through the campus of Warren Wilson College, and I imagined Stanley's ashes floating by while my daughters and I swam in the summertime. During my research, I'd seen how choices about our bodies after death are often influenced by a family's beliefs or traditions—or in this case, their business.

"On my mom's side, my grandparents chose burial," I said, "while my dad's parents were cremated, and they even installed a columbarium at their church in Mississippi, with niches carved into the walls of the stone to store the ashes." These structures were common in European churches, and I'd watched the priest place bags of their ashes into the stone walls of Trinity Episcopal Church in Hattiesburg.

"Family tradition even extends to which funeral home or crematory people call after a death," Stanley said.

Stanley told me the county had a budget for the cremation of bodies without a next of kin. They rotated through the eight different crematories in the area and paid $395 for each cremation.

"At the morgue, they will refrigerate the body for thirty days and try to find the next of kin," he said. "If they can't, then they assign a funeral home to do the cremation."

"So what happens when families ask to watch a cremation?" I nudged. By this point, I'd studied photos of the rectangular cremation furnace with an opening where the staff pushes the box into the heat, almost like a pizza oven. The crematories use a cardboard casket for the body, which goes into a furnace or cremation chamber, also called a retort.

"If a family wants to watch, we have them sign a waiver," he said, "because we don't know what kind of trauma they may experience from observing."

* * *

Since I didn't know anyone who would welcome my presence at a cremation, I turned to the next-best alternative to a lived experience: YouTube.

Before watching videos of bodies burning, I'd read accounts of cremation from funeral director Caitlin Doughty, whose book *When Smoke Gets in Our Eyes* details her work in a crematory, from picking up unclaimed bodies at the morgue to the heartbreaking task of cremating babies who had died. She describes leaving her job at the end of the day, when she is covered in "fine layers of people dust."

The story goes that the first coal-fired crematory in the United States was in Pennsylvania, where it was championed by a man named the Rev. Octavius B. Frothingham. After burning for about two and a half hours, the body disintegrated into ashes and bones, the same amount of time as it takes to cremate a body today.

Reading multiple accounts of cremation—from the shores of the Ganges to crematories in California—prepared me to take several deep breaths before searching YouTube for videos. The first link that popped up included bright-red text in large capital letters announcing, "WARNING!!!!" When I clicked the play button, a middle-aged man with an Irish accent reminded his audience that he'd warned us: this video was graphic and might contain disturbing images.

"I'm freaking glad I don't have this job of putting bodies into the furnace," the man in the video told his viewers. I wasn't clear about his actual work at the funeral home, beyond advancing graphic education for death literacy. But the images reminded me of a line from Caitlin Doughty: "Today, not being forced to see corpses is a privilege of the developed world," she writes.

In the video, I watched as two men placed a body wrapped in blue plastic into a cardboard box, taped the box shut, and then pushed the box into an opening in a metal oven about the size of a long kitchen counter, similar to yet longer than the ovens at Papa John's. I'd been told by funeral directors that if family chooses to watch, they only see the box going into the retort. Some families want to flip the switch that turns on the furnace, but the door is then closed.

This footage showed the flames with an open door to the furnace—followed by a pile of ashes surrounding a human skull and other bones. The staff took a long metal pole with a rake at the end and pushed around the white bones, much as you would stoke a fire with a stick. Finally, they raked out the remains, which would be ground into ashes in a machine called a cremulator. The image I couldn't get out of my mind was the white cranium with empty eye sockets, while the rest of the body burned to "people dust." As my daughters believe, maybe you can learn anything from YouTube.

"What are your wishes now for your body after death?" I asked Lyn, who continued to indulge my stories from this journey. "You've got us all talking about dying these days," she added.

As we chatted, I held the phone between my ear and shoulder while washing the supper dishes. Annie Sky danced between the bathroom and her bedroom as music streamed from her earbuds.

"I know we've talked about cremation before, but I'm learning even more about the impacts and the fossil fuels burned," I said, "The process emits mercury into the air and produces 250 to 350 pounds of carbon dioxide."

"This sounds like a lecture," Lyn said. "Have you thought about teaching this content in your classes?"

"Oh, I already have!" I said. "I'm still obsessed with that book called *Advice for Future Corpses*. The author writes that cremation of an adult requires 2 million BTUs an hour of energy. One gallon of gasoline gives 124,000 BTUs. So cremation emits greenhouse gases and other compounds, including the mercury from fillings. And yes, I'm reading from my note cards now."

"What's a BTU?" she asked.

"I had to look it up," I said. "The acronym stands for British thermal unit, and it's the amount of energy needed to raise one pound of water one degree Fahrenheit at sea level. It's just a measurement of energy," I read from the card.

"I still like the convenience of cremation, but you remember what happened when I lost my dad's ashes," Lyn said. "We'd buried most of his ashes in the VA cemetery, but I kept a small amount to scatter in Ireland with my kids."

"I'll never forget it," I said. "I still can't believe you used some of your mom's ashes as a substitute."

"I also burned random papers in the fireplace to add some volume," she added.

"Why didn't you just tell your boys?" I asked. "They were both in their twenties."

"I just couldn't," Lyn said. "I'd put those ashes somewhere for safekeeping but could not find them for the life of me. No pun intended."

"But you could tell me about it?" I asked. Raised by my father, an Eagle Scout, and my mother, the child of a priest, I was wired for honesty, even when it didn't serve me.

"Cremation is convenient if you don't misplace the ashes," she said. "End of story. Besides, I only needed a small pillbox for each of them to scatter the remains. Now when we die, the only option more affordable than cremation would be buying a plot in the Warren Wilson Cemetery for $100. But now you've told me they require vaults, so I'm not sure what to do."

"I'm working on the issue," I said. "Stay tuned. I'm drafting an email to the cemetery trustees this week."

"Cremation also responds to the issue of land scarcity," I continued. "That's why alternatives like human composting and water cremation are so exciting."

"Water cremation?" Lyn asked. "Is that cremation at the water's edge? Get it?"

"That's a good one," I said. "It's also called alkaline hydrolysis, aquamation, and wet cremation," I said. "The process uses heat, lye, and water to break down a body into liquid and the remaining bone, which can be ground up and given to the family. It's legal in twenty states, including North Carolina, and the number is growing."

"Are you reading from your lecture notes again?" Lyn asked before the conversation shifted to gossip about work and updates on our children.

I used to think my only choice for body disposition was burial versus cremation, but I'd learned about the growing popularity of alkaline hydrolysis, or aquamation—especially since its total carbon footprint is one-tenth that of flame cremation. I'd also read that alkaline hydrolysis produced 58 total kg of carbon dioxide, compared to the 243 total kg from cremation. In short, it appeared to be a more sustainable choice than flame cremation, especially for the climate.

The process occurs in a stainless-steel high-pressure cylinder filled with water and lye, heated to 200 to 300 degrees Fahrenheit. In six to twelve hours, the body dissolves, leaving the bones and the liquid, which could go down the drain. The costs are about $150 to $500 more than for flame cremation. In some states, casket makers had lobbied against legalization of aquamation for humans, due to the threat of competition.

One of the largest mortuaries in this area, Groce Funeral Home in Asheville, planned to invest in alkaline hydrolysis as a greener alternative to flame cremation. They hoped to purchase the equipment at a cost of about $200,000 and begin marketing within a year, predicting a demand for this more sustainable process. Alkaline hydrolysis isn't an environmental panacea at all, as the procedure uses 300 gallons of water, about three times the average daily use of someone in the United States. There are hidden costs as well: the energy required to produce four gallons of lye (potassium hydroxide) per cremation, the manufacture of the 3,000-pound stainless-steel machine, and the wastewater treatment needed for disposal of the liquid remains. But it still seemed like a better choice for the climate. I concluded that the sustainability ranking, lowest to highest, seemed to be conventional burial, flame cremation, alkaline hydrolysis, and green burial.

Funeral director Scott Groce came to my office one rainy February morning to share his plans for "AH," shorthand for alkaline hydrolysis. I'd offered to drive to his funeral home, which is close to Carolina Memorial Sanctuary, but he wanted to visit campus.

"This is the one day I don't have a funeral or a family visitation," he said. "It's good for me to get out of the office."

On a stormy morning with torrential downpours, he might have been the only person wearing a dark suit and tie on our campus, where students were more likely to don their Carhartt overalls and work boots than business attire in the rain, which hadn't stopped for two weeks. The Swannanoa River running through campus was close to cresting. But Scott accepted a hot mug of chamomile tea, and we talked in my office about this innovation in disposition of the body.

"If we educate people about AH, I don't think anyone will choose flame cremation," he said. "The environment is the number-one reason, but alkaline hydrolysis is gentler and quieter, and the family can be a part of the process. The aquamation centers are designed to include a room with a window where families can be present when the shrouded body enters the chamber."

As we visited, students came in and out of my office building, dropping off materials for faculty or running to work, unaware that I was talking with a funeral director who'd parked his black Suburban outside the building.

"Everybody thinks alkaline hydrolysis is the future of cremation," he said. "The main challenge for us is convincing my dad and uncle that the investment will pay off. But we want to have the machine in place and start the public relations about the process next year. We want to educate the media about the process as well."

The cheapest cremation offered by Groce was $2,800, and they would charge the same amount for aquamation.

"Cremation is more expensive with us than with low-cost crematories that can charge $1,000," Scott said. "We have more staff and nicer facilities, and we can provide more services, like a memorial. But aquamation will still be less expensive than a conventional burial with a vault and embalming, and it will be less harmful to the environment than flame cremation. The budget crematories won't be able to make the early investment in alkaline hydrolysis."

With three locations in our area, Groce Funeral Home expected alkaline hydrolysis to become popular with clients who had concerns about the environmental impacts of flame cremation or who wanted a "gentler" process. Some of the families served by Groce would always choose conventional burial, Scott said, but he expected aquamation to take off, given the rising popularity of cremation at a national level. In 2010, the Cremation Association of North America voted to expand the definition of cremation to include alkaline hydrolysis. In 2003, Minnesota became the first state to legalize aquamation for humans, and it's legal in twenty states as of 2020.

Because of the funeral home's proximity to Carolina Memorial Sanctuary, Scott had developed a collaboration with Cassie and Caroline, and Groce Funeral Home offered refrigeration of the body, the death certificate, and transport to the conservation cemetery for burial at a price of $1,410. He said this dialogue had "broken down walls" between them, and I could imagine how alkaline hydrolysis would also appeal to families interested in the sanctuary for environmental reasons. The ground-up bones from water cremation could be buried or scattered like ashes from flame cremation. Families could actually take the liquid—also called "essence"—and use it as fertilizer in a garden, due to the salts, sugars, and amino acids remaining. (Otherwise, the fluid could go down the drain.)

Until Groce Funeral Home offered alkaline hydrolysis, the nearest facility for the process was about two hours away in Shelby, North Carolina, at a funeral home that advertised the choice as a "gentle, eco-friendly alternative to flame cremation." This chemical process was originally developed to handle animal carcasses, with the first recorded use in the 1800s. While searching online, I found a website for an Asheville business offering pet aquamation. The process is legal for pets and costs about $150 to $400, compared with $100 for cremation.

When I shared this discovery with Lyn, she wanted to learn more before considering alkaline hydrolysis for her aging fat cat named Gigi, who had once belonged to her father.

During the summer of my search, I volunteered again at Carolina Memorial Sanctuary, eager to return to its meadows and woodlands and understand more about the efforts behind conservation and ecological restoration of the land. While my first experience there had been a shroud burial, my second involved a burial of cremated remains.

After I parked my car, I met Cassie and the bearded grave digger, Matt, who arrived in the parking lot in the golf cart. The August afternoon was overcast, cooler than the previous week, when temperatures had been hitting record highs, a reminder of the weather extremes we were facing.

"This service is for a woman named Carla," Cassie told me, "and we're expecting about twenty guests."

The first car to arrive was a white van driven by Carla's husband, DeWayne, with his son and two teenage daughters. Wearing a purple shirt and tan slacks, he slowly got out of the van, and his son carried a bouquet of red and white roses. After closing the door, DeWayne looked back, as if he'd neglected something important.

"Whoa, I almost forgot the ashes!" he cried. He then handed Cassie a cardboard box that fit in his two palms. His granddaughters jumped out of the van, their phones tucked into the back pockets of their shorts.

Friends and family had the opportunity to mix the ashes in a biodegradable urn with amendments—nutrients to neutralize the alkalinity of the ashes. Sometimes people wanted to have the hands-on experience of combining the ashes, soil, and amendments in the container right at the gravesite. But DeWayne had asked Cassie to prepare the ashes prior to the service, so she zipped off in the golf cart and took the box with her. A company called the Living Urn sells biodegradable urns complete with amendments, although Cassie questioned the necessity of the nutrients if the ashes are buried underground.

It's different aboveground. "You can't just scatter ashes on plants without amendments or they will die," she said.

It is critical to amend the ashes before scattering them, as the alkalinity can kill plants. In these cases, the staff at Carolina Memorial use a product called Let Your Love Grow, which is an amendment meant to be mixed with both soil and cremains.

I wondered about all the people I knew who threw ashes into outdoor spaces, assuming the remains of the bodies helped the soil. Others looked at different options for ashes. My friend Lucy had transformed some of her mother's ashes into a pendant she wore as jewelry. I'd also read about a company in Florida called Eternal Reefs that mixes ashes and cement to create reef balls sunk offshore, designed to be home for marine life.

"Many people want a tree to grow in the exact spot where they buried their ashes," said Cassie. "But if you bury ashes here, you have trees all around you."

In the gravel parking lot, two elderly women and a gentleman with a cane took advantage of the offer of a ride on the golf cart. Matt placed a small plastic stool on the ground to help them step up to their seats. As we started walking down the gravel trail to the site, I brought up the rear while Cassie walked behind the cart leading the processional. The crowd of Carla's sisters and their husbands all looked over seventy years old, except for the grown children and the teenagers.

At one burial site on our left, Carla's son stopped to point out a grave surrounded by trees and foliage. "Look, you can see where they recently buried someone," he said, assuming the demeanor of a tour guide. "See how they put the rocks around the perimeter of the grave. And it's important to make that mound so the dirt will settle in time."

This was environmental education in practice, I thought, with the participants becoming teachers along the way. In the back of the line, I greeted one of the younger family members, who looked about my age. "This is wonderful," she said. "We're getting all these older friends of Carla's to take a walk in the woods!"

When we arrived at the site, Cassie placed the funnel-shaped urn by the hole, which had been dug about two feet deep. To the side of the hole were two tarps filled with the dirt that had been dug up. The staff always replaced layers of soil in the same order in which they shoveled it out.

Later that day, I learned from DeWayne that his wife had suffered from cancer for seventeen years and had died a month earlier. She'd kept beating the odds, year after year, he told me.

"My sister had heard about natural burial, so I started looking around," he said. "I visited this cemetery three times—once with my son, once with the grandchildren, and then a final time by myself. We also looked at one of those hybrid cemeteries in town, but it just seemed like a manicured burial ground. We wanted something natural like this here."

After visiting the sanctuary, he'd looked through his wife's papers and found an article entitled "Green Reaper" about Caroline Yonge, the founder of this conservation cemetery. I'd read the same piece in a local women's magazine years ago.

"I was so glad to realize she would have approved of my choice," he told me.

In this low-key service, DeWayne had taken the lead but in a quiet and reserved way.

"This isn't gonna be fancy," he said. "It'll be real informal. But I want to invite any of you to say a word about Carla."

After this invitation, people started sharing remembrances, mostly short stories that involved travel. A grandchild recalled taking a cruise with them, while a friend remembered traveling in Romania together and being stopped at the border.

One of the elderly sisters, seated in a bamboo chair, talked about Carla's faith and how she kept a right relationship with God during the years of her illness.

Cassie picked up the urn and handed it to DeWayne, who placed the container in the ground. Then she invited anyone who wanted to shovel dirt to come forward and take a turn. Nearly everyone either picked up a scoop of dirt or shoveled some soil into the hole. Two men then reached for the shovels to close the grave.

DeWayne passed around the basket of roses, so people could place a flower on the grave. Each person stuck the stems in the ground, so the roses stood upright on top of the mound, like soldiers in battle or dancers on the stage.

Before we walked back, one of the elderly men looked at me and said, "Well, this is mighty different, but it sure is nice. This is how folks used to bury people."

Another cousin chimed in: "When my parents died, I buried the ashes myself in the cemetery. The manager told me to go right ahead but not to tell anyone!"

As we walked, I happened to fall in line next to DeWayne. "You did a nice job of holding space for all those people to talk," I said. By this point, I'd begun to use the vocabulary of the death and doula culture, with terms like "holding space."

"Well, I was okay as long as I didn't have to say anything," he said.

"I can't believe how many countries y'all have seen," I said.

"I worked for a cruise ship," he explained.

"How many countries did you visit again?" asked a relative ahead of us.

"One hundred," he said, "and that was for my job, mission work, and travel."

At that moment, I heard a rustle in the grasses by the trail, and a fawn jumped out and ran down the trail in front of us. We stood completely still and silent, watching her disappear.

When the cars pulled away, I joined Matt and Cassie on the golf cart to pick up the chairs and tarps from the burial site. As we headed back, we talked about cremation's popularity due to the cost and convenience.

"I couldn't be cremated, given what I know about the environmental costs," said Cassie. "But I think a lot of people will choose alkaline hydrolysis when it becomes more common. Our traditions, religion, culture, and family history all influence these choices."

After the experience at the conservation cemetery and my own exploration of cremation, I still saw this option as a plan B for me. In my father's two-page directives, he'd also specified cremation as an alternative if his death happened away from Alabama, while he was hiking some long-distance trail or kayaking a faraway river.

If alkaline hydrolysis were available when I died, that choice would make sense as an alternative, given the decreased environmental impacts. Aquamation reminded me of putting a body in a big Instant Pot for a few hours, much like a roast. Originally designed for domesticated animals, the process was good enough for a cow and also better for the environment: it would work for me as well.

Yet despite the people I'd met and the places I'd explored, one lingering question remained: Could I be buried at Warren Wilson without a vault? I'd fallen in love with the sanctuary, but the college cemetery close to my house still felt more like home.

As Cassie had said to me earlier, "The bigger issue isn't what you do with your body after death but how are you preparing for your death?"

I'd continued to reflect on a daily spiritual practice, inspired by my mother's meditations by candlelight at dusk and dawn and my daughter's intention to live with radiance.

"The number-one regret of the dying is that they didn't live a life true to themselves," Cassie had said to me. "How are we living our life right now? That's the biggest thing to ask in preparing for death. That paperwork lets others know how you want to die and what should happen to your body at the end. But first we have to ask ourselves how we are living."

Eight months into this exploration, I still wasn't sure how I was living, but I was definitely talking to my children and friends about death. By this point, however, the only time I could engage Annie Sky in these discussions was when she was captive in the car—an irony as I considered impacts of my death on the climate.

"I found out about a new way to cremate bodies with water," I told her as we drove home from her dance class. "It's called aquamation and uses water and high pressure to dissolve the body except for the bones."

"Yuck," she said, looking out the window.

"Burning the body could be considered gross too," I said. "But it's worse for the environment."

"You know I don't care that much, right?" she asked.

This wasn't the first time I'd heard that sentiment: I was not deterred in the least. After giving her a slight smile, I stared ahead at the road, feigning disengagement, my most powerful and least used parental trick.

After a short silence, she turned to face me and said, "Here's what I don't understand: If you have a place to bury your body close to home, why can't you just put your body back in the earth? It seems pretty simple to me."

Maybe the question wasn't existential at all. Perhaps this was exactly how we were living, side by side on the ride home.

Chapter 9

GIVING BACK

Body Donation, Decomposition, and Human Composting

When I peered into the cardboard box, I could identify two femur bones, seven ribs, a patella, and a skull, staring at me like a plastic skeleton at Halloween. The difference was that these remains came from a human body decomposed on a hill in the mountains of Western North Carolina, about an hour's drive from my house. There was no name on the box, only a label to identify the person: 2011–05, the fifth body donated to the lab that year.

"No one could have told me I'd be so emotionally impacted by the donors," said Dr. Katie Zejdlik. "Imagine the twenty-two-year-old whose parents decided to donate his body after he died so young." We were standing in the forensic anthropology lab, surrounded by boxes of bones, the remnants of the deceased donated to the "body farm" at Western Carolina University, known officially as the Forensic Osteology Research Station, or FOREST.

As we toured the lab, Katie explained that she'd directed the facility, as it's known, since she and her husband began work as forensic anthropologists at this regional university, tucked in the mountains between Asheville, North Carolina, and Knoxville, Tennessee. When I first met her, she was seven months pregnant while teaching full-time and administering the Western Carolina Human Identification Laboratory. This meant she conducted research and trained students in the recovery, storage, and analysis of human remains, including reconstructing skeletal remains and studying decomposition. I'd come to her office to see if body donation—where the body decomposes into soil—might be a sustainable choice for me.

As usual, I'd fielded the option with Annie Sky before driving to the lab.

"Wait, let me get this straight," my daughter said. "Your body would just rot on the ground?!"

I was starting to think that she might disapprove of any choice for disposition of my body, even after I explained my donation would benefit scientific research done by students.

"College students would study your naked body?" she continued. "No, thank you!"

This might actually be a teenager's version of hell on earth. I had to admit human decomposition appears "out there" at first glance. A friend told me it reminded her of *Invasion of the Body Snatchers* and the stuff of horror movies, but I actually didn't see much difference in decomposing aboveground rather than below it. When I first started this research, body donation and decomposition was a green end-of-life choice I hadn't even considered. One fall afternoon, I drove to Western Carolina University to find out more for myself.

"I never know who is going to call when I hear the phone ring in my office," Katie said as she showed me another box with bones tucked into labeled Ziploc bags. "It could be a grieving parent who wants to donate their child's body or a conspiracy theorist."

Because of the department's privacy policy, I wasn't able to observe the site where researchers like Katie and students study the process of human decomposition. But I'd secured an interview and visits to the lab, thanks to a phone call from the archeologist at my college, who vouched for me.

The site at Western Carolina University was the second body farm established in the country: the first one is located down I-40 West at the University of Tennessee in Knoxville. Altogether, the United States has eight such facilities, also called "decomp labs," where scientists study decomposition for diverse reasons, from seeking sustainable options for disposition to re-creating crime scenes for law enforcement. At this facility, the focus is the science of decomposition and skeletal analysis.

"Death does incredible things to the living," Katie told me.

She described her relationship with the next of kin, who have the right to see the remains—the actual bones used for teaching and research in the skeletal collection. The husband of a donor called her office once a week, like clockwork, to ask how his wife's skeleton was being used at that time.

"I think our connection was how he found meaning in his wife's death," Katie said.

From my online research, I could imagine the decomp facility with about ten bodies lying along a slope in the mountains as insects, vultures, and microorganisms did the beneficial work of breaking down human flesh. The outdoor site was surrounded by chain-link fencing covered in black fabric. My visuals came from YouTube videos documented before stricter privacy protocols came into play.

In 2017, *Esquire* magazine featured the story of a local woman, Kate Oberlin, whose bones were now in the collection at the lab. The family decided to document her death and ultimately the donation of her body.

"For the good and bad of it," she'd told the journalist, "there is a map."

After his wife died, Deloy Oberlin held a three-day home funeral for her, complete with music, friends, and food. The family then laid her body on the ground at the site to decompose.

When I was a child, my mother would place pieces of string on the ground to create four circles, giving each of her children a boundary to observe signs of life on the earth. I think she wanted us to consider the vast living world under our feet (and also get us outside). A single acre of soil contains 2,400 pounds of fungi, 1,500 pounds of bacteria, 900 pounds of earthworms, 890 pounds of arthropods and algae, and 133 pounds of protozoa. Both soil and a dead body are full of life. Given the action of bacteria, the resulting pressure breaks the skin, and the deceased becomes one with the soil.

"We are indirectly a sustainable and green option for body disposition," Katie said. "But that is not our reason for being like a conservation cemetery. We are here as a research facility, and there is no cost to donate a body, although people have to transport the donor to our lab. This appeals to people who don't want to finance the funeral industry and others who don't have the resources to do so."

As we talked, students came into the lab and began their work, cataloging bones and entering data into the computer.

"We're getting a lot of vultures this year," she added, almost as an afterthought. "So we're studying the impact of the vultures on decomposition."

I thought about how Lyn had seen concrete towers in India where followers of Jainism placed the dead in the open air for vultures to pick apart the flesh, leaving the skeleton. Early in my research, my next-door neighbor Morning had fantasized about a sky burial in Tibet; it turns out we can do the same right here in our Appalachian Mountains. My bet was that the romantic appeal of "exposure burial" might seem different behind a chain-link fence with black siding to keep out vandals and thrill seekers. But in the end, the outcome is the same: bones and some skin after organic decomposition, a death without harm to the earth.

"This is what we use if we need to clean up the bones," Katie said, pointing to the Crock Pots on a countertop besides the metal tables.

I tried to imagine putting an entire body in a slow cooker, but then I realized they used the cooking device to boil off remaining flesh from the bones, after the body had mostly decomposed.

When forensic anthropologist Bill Bass started the first body farm in Knoxville in 1987, people thought the entire operation was macabre, Katie explained, adding that TV shows have increased awareness of forensic anthropology.

"So *CSI* and *Bones* have helped the field!" I said.

If a person in this state wanted to donate their body to medical research, they typically signed up with the North Carolina Anatomical Board, and the body had to be healthy. In contrast, Katie got calls from people who didn't qualify for whole-body donations to science and were seeking alternatives. The facility could receive seventeen new donations a year. Unlike donations to medicine, the body isn't embalmed, as researchers at the body farm are studying organic decomposition, rather than preserving the body for further study.

Katie shared that there are no state or federal regulations about human decomposition, with the only statute being that humans need to be 300 feet from a water source.

"We will take any body," she said, "unless the decomposition is past a point or if the donor has a blood-borne disease."

They also don't accept bodies that weigh more than 250 pounds.

"We're on a slope," she said. "It's just not practical."

Katie envisioned a research site for collaborative science at the facility: "How does decomposition affect the environment, and how does the environment affect decomposition?" she asked. "We should be partnering with entomologists in addition to forensic anthropologists."

As we later visited in her office, I told Katie how I discovered the body farm at Western Carolina after reading *From Here to Eternity*, a memoir by Caitlin Doughty. As I followed her travels across the globe, in search of cross-cultural practices around death, I read about human decomposition on a body farm only an hour from my house.

To research her book, Doughty visited the facility at Western Carolina University with Katrina Spade, a researcher spearheading the Urban Death Project at the time. Spade is now internationally known for Recompose, the company she founded based on her innovative process to compost human bodies and generate soil as a by-product. Indeed, her research was one of the drivers behind Washington State's legalization of human composting. In 2019, Governor Jay Inslee signed the bill, and the law went into effect May 1, 2020. It's now also legal in Oregon and Colorado.

After a decade of planning, research, and fund-raising, the Recompose facility opened in a revitalized industrial neighborhood of Seattle called SoDo (south of downtown) in 2020. Human composting provides an earth-friendly alternative to cremation and conventional burial in urban centers with a dearth of green space. This revolution in death care produces about one cubic yard of soil—several wheelbarrows full—which families can take home or donate to regional conservation projects in need of compost. Composting happens when you mix materials rich in nitrogen, like food waste or a dead body, with items rich in carbon, such as wood chips. When you add moisture and oxygen, temperatures inside a compost pile can reach 150 degrees. The molecules can transform into rich soil.

The cost per person at the Seattle facility is $5,500, more expensive than basic cremation but less than a conventional burial and without the serious environmental impacts of both. During this monthlong process, nutrients in bodies after death support life, saving an estimated one metric ton of carbon per person over standard burial or cremation.

"I've spoken to Katrina Spade," Katie said. "We're studying decomposition here, rather than composting. But animals in agriculture are composted all the time, and if human composting takes off, it could be phenomenal."

"People think there are only two options—burial or cremation," she said. "Cremation has outpaced burial, but I think we will see an uptick in alternatives for disposition like composting. I hate that word, *dispose*," she added, "but I don't know what other one to use."

"It's funny, because when I think of disposition," I said, "I think about someone's personality or their essence. Maybe our disposition while we're alive can influence the disposition of our body after death."

While the facility at Western Carolina doesn't focus on human composting, I was curious why people decide to donate, especially when body decomposition somehow seems more graphic aboveground than below it.

"People want to do something sustainable that gives back," she said. "At the decomp facility, there is a legacy. You donate your body to research here, and then as a skeleton, your body becomes a teaching tool. If you donate to a med school, you just get cremated afterward."

"Ultimately, after your death, you are decomposing," she said. "We lay your naked body on the ground in the mountains of Western North Carolina and study the process."

"Oh my God," I said, "It seems like every outdoor-loving person I've talked to about their final wishes wants someone to lay their body in the woods. They could actually do it!"

About half of the donors at the facility are what they call "pre-donors," as they've completed the form online in advance of their death. But the lab also accepts donations from next of kin who decide to donate their loved one's body after they die.

"We also get individuals who don't have money," she said, "because all you have to do is get the body here."

As she stood up from her office chair, Katie explained the university was building student housing right next to the facility, and the department was going to have to move the entire body farm.

"Where are y'all moving it?" I asked.

"Well, it's not clear yet," she said. "That decision is in the works."

I didn't want to ask too many questions about the logistics of moving a body farm. It had to be more complicated than renting a U-Haul for furniture, which seemed easy compared with moving bone fragments and soil particles.

Katie could tell what I was thinking: "I know," she said. "At least we can collect some good data by starting at a new site. I'm looking at it as a research opportunity."

"We were expecting a donor at one o'clock, and it's already four," said Christine Bailey, who offhandedly mentioned she was wearing her field clothes underneath the long, flowing skirt she threw on for meetings.

"There was no time to change," she said to me. "It's been that kind of day."

Professor Bailey, as she was known to the students, sat down on a silver metal stool across from me in the human osteology lab, where I'd returned several weeks after my first visit with Katie. Osteology is the scientific study of bones, a subdiscipline of anatomy, anthropology, and paleontology. Was I ready to donate my body to the body farm? I wasn't sure. First, I wanted to spend some time with the students engaged in hands-on learning at this lab.

Her work-study student Wesley was using a toothbrush to clean the dirt off the scapula of another donor. Seated on a stool at the autopsy table, he dipped each bone into soapy water, scraped off dirt with forceps, peeled off remaining flesh with tweezers, and then rinsed the bone in clear water. The process felt methodical, practical, and respectful, by virtue of the care he took with each bone. Wearing a white lab coat and safety glasses, he focused on the task in front of him while I listened to his supervisor.

While her colleague Dr. Z—Katie—served as the administrative director, Professor Bailey, the curator of the forensic anthropology facility and the lab, was the "boots on the ground" supervisor, overseeing the thirty-five student volunteers selected from a pool of eighty applicants, as well as the work-study students, the lab, and the donation program. She'd applied for the job after getting her undergrad and master's degrees in biological anthropology with a focus on forensics at the University of Tennessee–Knoxville.

"This job felt like it was made for me," she said. "There aren't a lot of people who've been the volunteer coordinator at a body farm."

In addition to overseeing the students, she was in contact with families of donors, funeral homes, and transit services. At the facility, she worked in skeletal recovery with her team of students, training them to do photography, mapping, and maintenance of the site.

She said the facility at UT-Knoxville had tried to avoid using the phrase *body farm* for years, but now it had embraced the term as its brand. "If the people who donate bodies are calling it the body farm, and the community calls it the body farm, why wouldn't we just do the same?"

Through the outdoor facility and the indoor lab, she and Dr. Z were training students for jobs in academia, positions as death investigators with law enforcement, and careers in biology and anatomy. As I watched Wesley's serious expression, I could imagine continuing my life as an educator—after my death—by donating my body.

Since I couldn't go to the actual body farm, I was hoping to witness a body donation, and this lab visit could be my chance. Whereas UT-Knoxville received eighty bodies a year, the number at Western Carolina was much smaller, about fifteen to seventeen bodies. I'd been in the lab for three hours by that point, but within minutes, Professor Bailey's phone rang.

"Okay, they are thirty minutes away," she said after hanging up.

"Would it be appropriate for me to stay?" I asked, trying not to sound desperate. "I'll keep out of the way and just observe."

"I don't see any problem, but let me check with Dr. Z," she said, sending off a quick text.

"Okay, she says it's fine," she said seconds later. "But they found the body several days ago, so there will be a smell. Just leave the room if it gets to you."

Silently, I vowed not to leave the room. In Mary Roach's book *Stiff: The Curious Life of Human Cadavers*, I'd read her description of the smell of decomposing humans: "It is dense and cloying, sweet but not flower-sweet, halfway between rotting fruit and rotting meat." Some part of me wanted to be in that space with the immediacy of dying. I'd been researching options after death, and I wanted to get closer—in space and time—to the end of our lives.

Another student had joined Wesley as he continued to process the bones. As soon as Professor Bailey left, they both told me this was their first time to observe the arrival of a body donation. Wesley opened a drawer and pulled out two surgical masks for him and his friend.

"Do you want one?" he asked me. "I don't know how bad it will smell."

While I took the mask, I later learned face coverings provide protection from body fluids, rather than odors. None of us in the lab that afternoon realized that in months to come, the COVID-19 pandemic would make masks a part of our everyday lives.

"I think they're here," Wesley said after we glimpsed Professor Bailey walking at a fast clip down the hall.

Soon she ran back into the lab and announced, "He's leaking. Put the liners in the cooler."

In response, Wesley ran to put absorbent liners into the cadaver cooler, used for keeping the bodies cool until they could be transported to the outdoor facility. It looked like a human-size toaster oven with pull-out trays that could hold bodies.

Wesley and I held open the door to the lab while Professor Bailey helped push the gurney along with two women who'd driven the car for Mountain Transport Services.

A thick blue body bag lay on top of the gurney, which they wheeled into the lab. I could see bloodstained fluid dripping out of the bag and onto the floor. This was the first time in my life I'd smelled the distinct odor of decomposing human flesh: rank, pungent, overwhelming but oddly familiar, as if it could have been me.

After pulling the gurney up to the cooler, the two students, wearing plastic gloves and sleeves, stood by one side of the body bag while the women from the transport company and Professor Bailey got on the other side. They strategized about how to lift the body and transfer it quickly to the cooler to prevent spillage. But when they lifted, the weight was too heavy. I saw my chance to help and ran to get my own gloves to avoid contamination with any leakage.

"I'm right here!" I called out, overeager to assist.

"Yes, come help us out," Professor Bailey said, "but make sure to put on the plastic sleeves too."

I stood at the front of the gurney, and we lifted together, sliding the body into the cooler and closing the door.

"Can I wash off my gurney here?" asked Jane, one of the women from the transit company. "And do y'all have a mop?" This was just another afternoon in the body transport business, an enterprise I hadn't even heard of until today.

Wesley took to the job of mopping up the bloody fluid on the floor.

After a long day of driving and waiting, Jane was just now catching her breath.

"We waited two hours at the hospital for a body bag that wouldn't leak," she said, holding her right hand to her head and pulling the pointer-finger trigger to signify the cause of death. Everyone got quiet for a moment.

"I'm going to donate my body to science too, but maybe to a medical school," she said, returning to the topic at hand. "I may have to take out the carpet in our vehicle. That smell won't be easy to remove."

The students washed off the plastic covering of the gurney, and Professor Bailey, as if anticipating a need, recommended a barbeque place along the driver's route out of town. The whole process felt so matter-of-fact but deeply respectful.

"It's not like you can stop and get a meal with a body in the back of the car," Professor Bailey told me. "I've learned that the staff from the funeral homes and transportation services are hungry and tired by the time they get here."

Before the women left, I asked the cost of transporting a body from Asheville, sixty miles away, to the lab. Professor Bailey had told me the lab had begun requiring families to use a funeral home or transit service for transportation.

"We charge $2 per mile," Jane said, "And we only get paid for miles when the body is in the vehicle, so we are driving home on our own dime."

If I went with their transport company, I'd want my daughters to include a generous tip. Professor Bailey later told me that some donors prepay for transportation, and the price from Asheville ranges from $150 to $500, depending on the company. She often recommended Asheville Mortuary Service for their reasonable rates.

As the students finished cleaning the lab, Professor Bailey checked her watch, shifting to the logistics of a late Friday afternoon, "I'm glad they got here in time," she said. "I've got to dog-sit in thirty minutes."

But she took a moment to reflect on the work of the afternoon: "When we're around death every day, it doesn't mean that we don't respect it. Instead, death just becomes a part of daily life."

"Your students were so thoughtful and professional throughout the donation," I told her before she left.

"They want to be here," she said. "That makes a difference."

I could imagine wanting to be here too, even if it was hard to visualize vultures pecking at my eyelids. It wasn't that different from bacteria decomposing my flesh underground. Perhaps that impression would change if I were able to actually see the facility, with human bones protruding from wood chips. But I felt at home in this learning environment, where students were discovering who they were by what they studied with caring mentors by their side.

Before the donation, I'd spent time in the lab with Wesley as he compiled a donor file on the computer. Individuals can predonate online by completing the forms, or family can arrange the donation soon after death. On the table behind the office, I saw reference books including *Human Osteology* and *Human Bone Manual*. There were labeled drawers with bags in a variety of sizes for collecting remains after decomposition.

With a caring look, Wesley had told me we'd begin processing a donor, but I'd end up spending nearly three hours watching him work before the body arrived at the lab. In another room, filled with cardboard boxes of bones, he pulled off the lid to a blue plastic bin and followed a protocol of reassembling the skeleton according to its anatomical placement.

"This is 2019–6," he said, "So it was the sixth body donated this year, and we recovered the bones on August 30. The body was likely donated sometime in April, so it took about four to five months to completely decompose."

After he picked up two scapulae, Wesley explained, "He just looks dirty, not fleshy. I'll wash each bone."

The protocol was listed on the wall, with photos showing students exactly how to re-create the skeletons and package them in the boxes.

"So here's the two scapulae, the ossa coxae, and the sacrum," he said, showing me bolts in the vertebrae that looked like a carpentry project at Home Depot. "Looks like the lumbar vertebrae were bolted during a surgery."

Next, he pulled out the cranium, which was cut, almost like a woodworking project.

"The cut indicates the person had an autopsy, since they pull out the brain after cutting the skull."

As Wesley reassembled the bones, I remembered writer Sallie Tisdale's reflection as she held the bones of her Buddhist teacher after his cremation: "We are a loose collection glued briefly into a provisional thing called self, and all such things are bound to dissolution."

While he worked, we chatted about his path to this job sorting bones from a body donated to science.

"Well, I'm from Nashville, North Carolina," he explained. "Yes, there's a Nashville in this state. And I thought I wanted to study engineering, but I found out my first semester that it wasn't for me. Now, don't tell Dr. Z, my advisor, but I just found a few different majors, rolled a die, and it landed on forensic anthropology. It turns out that I'm good at it!"

"You're only a sophomore, and already you're doing such impressive work," I said.

Wesley looked up from the bones with a smile. "Yeah, I'm kind of impressed with myself too. I've learned so much on the job."

He laid the skeletal parts on what looked like lunch trays and carried them to the autopsy table, where he would clean them using tools like soap and water, forceps, tweezers, toothbrushes, picks, bottle brushes, and scissors. Wesley had decided he wanted to go to graduate school in biological anthropology, just like his mentor, Dr. Z.

"A raccoon can decompose on the road, and no one freaks out," Wesley said. "But we have a messed-up attitude about human death in our society. Embalming is one of the worst things in our culture, along with our avoidance of death. There is something wrong about injecting your dead with chemicals and putting the body in a concrete vault."

"Did you know that research from this lab was used to help legalize human composting in Washington State?" he asked.

I nodded. We both speculated there would be a growing interest in and spread of human composting—pun intended—as a sustainable choice after death.

That afternoon at the lab, I also spoke with Anna, a senior from Germany, who planned to use her degree for graduate school in archeology and field recovery in the United Kingdom.

"I've wanted to study forensic anthropology since I was in high school," she said.

Anna had traveled to Germany with a team of anthropologists led by Dr. Z's husband, Dr. P. (I still had to pause to remember which letter referred to which professor.) In Europe, the research team had worked to find remains of missing soldiers from World War II.

"The first time I went to the facility was in an intro class," she said, "And the experience shows you if you can handle seeing a dead body and the smells. But once you're there, you realize this is a person. Someone loved this person."

She was on a Friday crew that went out every week to the facility.

"I wasn't scared to see a dead body," she said. "I'm so impressed by the process, and I've learned even more about osteology from working in the facility than from lectures in classes."

In a hoodie, jeans, and Converse shoes, Anna pulled her blond hair into a ponytail as we continued to talk.

"How do you recover the skeletons at the site?" I asked.

"We lay out the bones in their anatomical positions on tarps and then use a screen to find the small bones, like the wrist bones," she said. "I've recovered the bones and done inventory five or six times this semester."

I noticed a cranium in the box labeled 2018–06 in front of us. I wondered how working here had influenced Anna's attitude toward body donation. These students, with their curious natures and honest care about the world, reminded me of my own.

"It makes sense to donate your body to science," she said. "As students, we can learn so much from seeing the actual process."

"What about if your parents donated their bodies?" I asked.

"If I thought about putting my parents in the facility, it would break me a bit," she said. "But their donation would be of value to other people. Honestly, I would be more freaked out to have their ashes in my closet."

Unlike Jane with the transport service, whose family members had donated their bodies to medical science, I'd never known anyone in my family to make that choice. Many medical schools, such as Wake Forest University in North Carolina, accept bodies for donation, with the deceased or family members covering the transport costs. After embalming the body and conducting research, the medical school would then return the ashes to the next of kin; the average wait time was about one year.

When I brought up the option of body donation with my book club, I learned that one of our members, a pragmatic physical therapist named Debbie, had already completed the paperwork.

"I filled out the online forms years ago and got my sister as a witness," she said. "I'm donating to the Genesis Legacy program, which arranges and pays for transportation to their site in Memphis and then cremates your body. Your family contacts the program at the time of death for the transport. It's that simple."

"Does anyone want more wine?" she asked, moving on to the next order of the evening.

It didn't surprise me that someone who worked in the medical field would make these arrangements. In my research, I found that many people were aware of the financial and logistical convenience of "donating their body to science," but they didn't know about the option to donate their body and promote sustainability for the earth. That was the difference between the body farm at Western North Carolina and a medical school, which embalms and then cremates the deceased. This distinction seems important for anyone who considers giving their body to research.

Because one of my first criteria for disposition was to avoid embalming, donation to a medical school didn't meet my goals for sustainability and the climate. The program at Western Carolina University was earth-friendly and affordable, but I would have to reconcile myself to decomposing above the ground rather than below it. I'd read that Buddhist monks meditate on the image of a rotting corpse—the stages of decomposition—to prepare themselves for death. The stages include distention, rupture, exudation of blood, putrefaction, discoloration and desiccation, consumption by animals and birds, dismemberment, bones, and dust.

I felt like my meditation on death, in contrast to that of the monks, was the continuing dialogue with my two daughters about disposition. While we were driving to dance classes or chatting on the phone, we had casual conversations like the ones my father had with me when he built the tiny prototype of his wooden casket.

After spending time in the lab, I could imagine my body on the ground while Professor Bailey directed her students to take photos or reassemble my bones on a tarp. Maybe Wesley would even get his PhD and return to work in the lab, explaining to students how to reassemble my bones after decomp. I'd spent my career encouraging my students to connect to nature in our classes and in the field. Hanging out with Wesley and Anna as they talked about their work felt like comfortable terrain to me, a place I'd been for my entire adult life.

During this research, I also read about similar options, including a technique called promession, which turns bodies into composted soil through a process of freezing. Legal in Sweden and Switzerland, this process puts carbon from our bodies back into the soil, rather than releasing it through cremation. With every step of the way, I saw additional choices for disposition beyond those I'd imagined months before.

Donating a body for decomposition only an hour's drive from my house seemed so practical, and the site had been a part of the research behind human composting in Washington State that could expand to other parts of the country. Decomposition is one of the most basic metaphors for endings and resurrection, the way death can bring new life. But when I tried to talk to my youngest daughter about this option again, she shut it down.

"No, no, no, no, and no," she said after I picked her up from a friend's house, holding up her right hand to signal the end of the conversation.

"So what is the problem with body donation so close to home?" I pressed. "Can you describe it to me?"

"I'm tired of you asking my opinion about death," she said.

"Okay, that's fair," I said, stopping at a red light. "Is there another reason?"

She paused for a second and reached for the radio.

"It's simple," she said. "I want a place where I can take my children to visit you."

"That's what you want?" I asked.

She nodded and turned up the radio to Lizzo's song "Good as Hell," followed by Ariana Grande's "God Is a Woman." In the middle of everyday life, we meditated on death, one short conversation at a time.

Chapter 10

WRITE IT DOWN AND TALK ABOUT IT

Planning Directives in Extraordinary Times

"I WANTED TO LEARN more about being a friend with death," an end-of-life doula told me in a crowded coffee shop when I asked why she'd pursued this work.

It was early 2020, months before COVID-19 closed cafés and classrooms. Since then, I'd thought a lot about what she had said: imagining end of life as an unconventional BFF or soul mate. My father's directives had grounded me even as his death leveled me. I'd wanted to provide the same road map for my children, with both the climate and community in mind. At the end of the year, I considered my own end-of-life plans in the midst of a global pandemic, which left bodies stacked in refrigerated trucks in urban centers and turned gatherings like indoor church services into a possible death sentence.

Weeks before quarantine and social distancing, I stared at a pile of papers spread across my kitchen table, the same files I'd received at the beginning of this journey from the Center for End of Life Transitions. The paperwork included advance care directives and after-death directives: my death certificate, the Five Wishes living will, a dementia directive, the health care power of attorney, do-not-resuscitate (DNR) orders, and a declaration regarding disposition of the body.

Friends now talked to me about death without prompting and tagged me in relevant news stories every week. In this way, I learned about a nonprofit called Narrow Ridge Center in East Tennessee, which set aside five acres for free natural burial: no casket, no embalming, no cost. If the family couldn't dig the grave themselves, they could hire a grave digger for $250 as the only expense. Grassroots innovations reclaiming death traditions, especially in this Appalachian region, reinforced my decision to bury my body close to home without negative impacts on the land.

During the year, I'd revised my advance directives—the instructions for my care leading up to death. Three years after my sister's cross-country move, I finally added the names of my older daughter and Lyn as alternates in case my sister couldn't perform the duties of power of attorney. They would serve as my decision-making team. In my thirties, when I first completed a living will, I hadn't wanted extreme medical interventions to keep me alive; I now added the provision that I didn't want my life prolonged in the case of dementia. I'd watched my parents struggle to care for two of my grandparents when their bodies worked but their minds didn't. For us all, these documents were like relationships requiring updates to maintain their relevance.

This journey to leave the earth without harming it revealed diverse options: home funerals, conservation burial, natural burial, cremation, alkaline hydrolysis, and body donation. The decision about disposition felt more complicated than the advance directives, even after one year of research.

Despite months of conversation and emails, I still hadn't heard if I could be buried without a vault at the Warren Wilson Cemetery. It seemed to me that Ray Stock, the eighty-five-year old cemetery trustee, remained the primary gatekeeper to a green burial at the college.

Months before the pandemic, I'd arranged to take my environmental education class to the conservation cemetery where I'd volunteered, Carolina Memorial Sanctuary, which remained my other choice for a natural burial. Several of my students at the time were active in the Friday climate strikes, pushing municipalities across the United States to act on climate policy. In response, the city of Asheville declared a climate emergency and announced concrete steps to address it. Around the same time, a federal appeals court ruled against my former student Kelsey Juliana, the lead plaintiff among twenty-one young people suing the government to demand a plan to confront the climate crisis.

My research and teaching left me with questions: Why did youth have to fight so hard for their right to a healthy life? And how could my one death connect to the existential crisis facing the planet? I understood that my students, children, the climate crisis, and death intersect in my life, even though I didn't understand how all the threads relate. We were all bound to each other, in ways both seen and unseen—a lesson we'd take to heart when quarantined inside our homes and later protesting for racial justice in the streets.

I wanted the students in my class to think critically about how death isn't only a personal loss: the ways we take our leave of this earth have consequences in communities. On a soggy, cold afternoon, we walked slowly past the mounded graves at Carolina Memorial Sanctuary, as alum and manager Cassie Barrett shared stories of the dead and updates on the ecological restoration of the wetland around us. Given the rain, I scanned the group for signs of hypothermia, just in case. As we gathered in a circle at the end of the tour, Cassie asked the students one question: "Was there anything uncomfortable about death that came up for you?"

The students looked down at their feet, their shoulders hunched over in the chill. Someone finally responded, "In my family, we didn't talk about death much at all, so today has been the most I've spoken about dying in a long, long time."

Cassie nodded. "We're not taught to talk about death," she said.

A young woman studying forestry said she'd never thought about the role of conservation at cemeteries: "I didn't know burials had anything to do with the environment."

Although they were somber at the cemetery, the students had plenty to say on the final exam when asked to explore the connection between death and the future of the earth.

"My mother, who has multiple autoimmune diseases, speaks about assisted suicide," wrote one student. "I tell her all the time I don't want to discuss it, but I think when I go home, I will finally sit down and talk with her."

She continued: "I was taken aback at how different this cemetery was. It didn't feel like this packed, claustrophobic, liminal space of death. This to me is what I consider a final resting place, one in which family members could visit and take a therapeutic hike on the trails."

Other students were equally affected. "I lost two really important people since coming to college," one wrote. "I wished that I could have buried my grandfather and my best friend in a similar place as this. It turns something so painful into something beautiful and alive. It broadened my understanding of how important the land can be, in this case as a way to honor the deceased."

I couldn't complete my death plan without knowing where I wanted to be buried. I'd fallen in love with the vision at Carolina Memorial Sanctuary, but the small cemetery at Warren Wilson College was home for me and my daughters. When I last met with Ray, I'd offered to cook dinner for all three men who served as cemetery trustees if they could entertain the option to forgo vaults. Rev. Steve Runholt suggested a more direct approach: "Write an email asking for an exemption to the burial contract based on religious reasons," he said.

This was a novel but authentic strategy: request a change based on shared values.

"Can I copy you?" I asked. He hesitated and then nodded.

So I drafted my message: "I am asking for a one-time exemption to the requirement for a vault, as my Episcopal faith and the climate crisis propel me toward a green burial to care for the earth." I also noted the exemption for the student Sheriff Disu due to his faith, reminding them that Muslim and Jewish traditions prohibit the use of vaults. During that burial, the imam had stepped into the grave to turn the young body in these Appalachian Mountains toward Mecca.

The next day, I got a promising response from Bill Sanderson, a retired high school science teacher and new cemetery trustee. While he didn't know the history and legal issues surrounding vaults, he looked forward to "working for a creative, and greener future for the cemetery."

Several weeks later, I still hadn't heard from the other two trustees.

"If Ray doesn't respond, there's not much I can do," I wrote to Steve. "Or is there?"

"We're just going to have to wait and see," he said.

This strategy was not my strong suit. It hadn't been my father's either.

This year, I learned many lessons that surprised me about how we choose to take our place in the world after we die, the environmental upshot of that, and the surprising ways people can lessen the impact of their death on the earth. Just as it's important to talk about the climate emergency, since most people don't hear friends and family discuss it, we must also talk about death in everyday life. The two are deeply connected.

The more comfortable I became talking about death, the more the topic seemed to become a part of everyday conversation. After a faculty meeting one day, the Appalachian Studies professor approached me and said, "If you want to talk about death now, we can." He was descended from five generations of tombstone salesmen in Kentucky and had mowed cemeteries to make money in the summer. I'd worked with him for ten years without knowing this family history. As we spoke, he reminded me of the strong tradition of home funerals in this region.

Most importantly, I knew it was time to talk with my daughters. They were the audience for my directives, the ones who would feel the greatest impact of my death. Their father and I had lived apart since Maya was four years old. If I died, their lives—especially Annie Sky's—would change in irrevocable ways. I'd planned to take them both to the conservation burial ground and the Warren Wilson Cemetery soon after spring break, when we learned Maya's college was closing due to COVID-19. She had two days to pack up and return to our small home, where she hadn't lived full-time in three years.

Maya moved back into her room, a walk-in closet, sharing a bunk bed with her sister and one tiny bathroom with us all. During that same frenetic weekend, I shifted my experiential courses into some semblance of virtual learning on Zoom. Annie Sky set her morning alarm to start school on her laptop and waited until the night to ask me, "Are you going to die from corona? What happens if you die?"

The questions I'd been asking all year seemed to matter both less and more at the same time, especially as she watched CNN Student News for social studies and saw the death rate climb in the online graphics. There was a sense of collective loss as we pecked at our screens while spring paraded bright colors with white dogwoods, yellow daffodils, and redbuds on campus.

During the pandemic, educators talked a lot about a "flipped classroom," in which students are introduced to material online and later engage with the content in class. From my window on campus, I saw a "flipped spring," during which the bluebirds sang with irreverent joy but we humans were sequestered inside on our devices. Rachel Carson wrote about a silent spring caused by the ravages of DDT on wildlife, but the coronavirus attacked its human host. It silenced our lives in so many ways. At night, I used my spray bottle of bleach and water on cabinet handles, toilets, bathtub knobs, and doors, erasing a threat I couldn't see.

Early during the first spring of quarantine, I got a text during a weekend when my girls were at their dad's house: "Maya has coronavirus," Annie Sky wrote. "She's coughing, and her chest hurts. I know you'll say she doesn't have it, but she does."

She was sick for more than a week, barely leaving the top bunk of her bed. Though I often have a tough-love attitude toward illness, another inheritance from my father, I stared at her matted hair on her pillow and wondered, "Could a twenty-one-year-old die when I'm gone for a run?" After ten days or so, she slowly got stronger, yet there was no available testing at the time, since she didn't have a fever. We all faced mortality in the midst of a life-giving spring. Maya began to stay up late at night, doing homework online with friends, and then sleep during the day, especially after hearing about the cancellation of her summer job. It was like she'd gone on a study-abroad trip in a different time zone within our tiny home.

After teaching an online class from my bedroom, I received an email from a well-known theologian and author of more than ten books, including ones on creation and the Bible. It was early in the pandemic, during the season of Lent, marking the forty days before the death and resurrection of Jesus.

"In this season when for multiple reasons we ponder our mortality with earnestness, I am realizing it is time to make concrete plans for death," she wrote. "My husband favors cremation, but we are also aware of the environmental costs. We are seeking wisdom on this."

My friend Brian had connected us, and I was starstruck, given my intellectual crush on this thoughtful writer. After texting Brian to gush, I wrote her back and shared options for natural burial near her home, as well as the choice of aquamation, which uses less fossil fuel than flame cremation. Within a few hours, she'd spoken to staff at a facility called EternaCare, which offers alkaline hydrolysis, including transportation, for only a few hundred dollars more than cremation. She soon finalized plans for herself and her husband, who was in his nineties.

"Yesterday I awakened with the recognition that I had no idea what either one of us would do at the moment of death, and now I do know," she wrote. "This is a significant relief."

With one phone call and some online searching, she'd found a choice for her body that reflected her life's work for this good earth—during a pandemic, when the time was right to plan.

She was not alone: More people were documenting their end-of-life wishes, some using online services with pithy names like "Cake" and "Lantern," which saw more than a 100 percent increase in users during the pandemic, most under the age of forty-five.

Many states limited attendance at funerals to ten people and also required masks and social distancing. Funeral director Amy Cunningham, who founded a green funeral home in Brooklyn, had to store caskets in her living room due to a shortage during the pandemic. While handling the overwhelming number of bodies, she created small rituals like playing music in the hearse. Funeral directors and clergy became on-the-ground innovators: At Larkspur Conservation, the staff gave each person a color-coded shovel to ensure that those gathered could participate safely in closing the grave. The Rev. Becca Stevens, cofounder of this conservation cemetery in Tennessee, spoke about the pandemic in Nashville: "This is a new experience for our community, but it's not a new experience in history. . . . Knowing love transcends six feet and generosity can easily span the distances we have to keep."

In quarantine, we stood six feet apart, and we often bury six feet under. But we have the opportunity to move toward a world that cares for both people and places. Climate scientist Katharine Hayhoe drew parallels between this pandemic, the climate, and life and death: "What really matters is the same for all of us. It's the health and safety of our friends, our family, our loved ones, our communities, our cities, and our country," she wrote in an Instagram post. "That's what the coronavirus threatens and that's exactly what climate change does too."

During the early months of the pandemic, I began to worry less about my classes and more about the mental health of my small family in our 900-square-foot home. Even though we had everything we needed, I hadn't felt this vulnerable since Annie Sky was a newborn and I was alone as a parent. The manila folder in my room still held the form for disposition of my body, the end of this journey in the middle of the pandemic.

The question of what would happen to my body when I died became overtaken by my uncertainty: How do I live, teach, and parent with children isolated and online in my home? After my virtual classes each day, I'd bike with Annie Sky to the farm where a core crew of students managed the animals, including baby pigs lolling in the sun while country music blasted from speakers into their pens.

"I think they seem happier with the music, don't you?" asked my fourteen-year old. "They don't care about corona."

One weekend, a neighbor furloughed from his landscaping job volunteered to put in a garden for me. This was a man who grew most of his food in a tiny space in front of his home on campus.

"Something's gonna shift, I tell you," he said while placing tiny seeds into the soil. "Things can't keep growing and expanding without people taking care of each other and the places where they live."

This spirit reminded me of writer Rebecca Solnit's line: "Disasters often unfold like revolutions."

As the seeds germinated in my garden, I saw even more clearly that how we decide to live after the pandemic connects to decisions we make for our bodies after our deaths. There was much at stake for individuals, the collective, and the earth. These interconnections became even more real as I joined community members in the streets that summer with masks on our faces and cardboard signs in our hands that read, "#BlackLivesMatter" and "No Justice, No Peace."

Standing at a prayer vigil in Asheville, I heard the mayor break down in tears after another night of clashes on the streets between police and protesters. Across small towns and big cities, people showed up to demand justice in the face of the murders of Black Americans at the hands of police. The city of Asheville and the surrounding county voted to support community reparations and take down monuments to white supremacists. One outcome of the protests and the pandemic was that we were all thinking about death at the same time.

I'd thought this year of research would bring a crystal-clear answer on a small scale: an unveiling of a green future at the Warren Wilson Cemetery. In my daydreams, I also imagined the year would bring me closer to living like the end-of-life doulas I'd met or my own mother, who wove spiritual practices into her days. The truth was that during the pandemic, I was just hanging on, day by day, holding together our little household in some way.

One afternoon, I cleared the top of a bookshelf in my bedroom and placed on it a candle in a glass container etched with the words *Love Heals Every Body.* Next to the candle, I put some essential oils and a taller Virgin Mary candle purchased from Dollar General down the road. Without thinking, I'd made myself an altar. It wasn't a transformed life, but it was a call to stillness, a silent reminder that we carry inside us the love of those who have died—perhaps the spiritual practice I actually needed. A messy ending quarantined with my daughters was the pause in the story, a narrative stitched together in a painful, fragmented world.

As a self-proclaimed climate feminist with two daughters, I didn't intend for this story to close on a Zoom call with four men, but that's what happened.

"Sometimes older folks have trouble on Zoom," Steve had warned me in an email before our virtual meeting with the trustees.

"You've got quite a man cave there, Bill," said Steve, noting the compound bows, antlers, guns, and books on the wall behind the retired teacher. His wife worked as the archivist in the college library.

"Here comes John," said Bill, as a black box appeared in the screen.

"We can't see you, John, but that's okay, as long as you can hear us!" said Steve.

"I can hear every word!" he said.

"Well, Ray isn't here, but we should move forward without him," Steve said. I couldn't envision the committee making a decision without their stalwart leader, who had the most sweat equity in the land.

"Wait, here's Ray!" said Steve. We could see him but couldn't hear him.

"Unmute your mic!" said Steve. I felt gut-wrenching flashbacks to my first week of online teaching that spring.

After Steve convened the meeting, he asked me to summarize my request and then called for questions. Many of the concerns focused on maintenance of the site without a vault.

"Who's going to maintain your plot if the ground sinks?" asked John. In response, I described adding dirt every few years to the grave sites of my parents. Perhaps we could include a maintenance fee for green burial, I suggested, or a clause in the contract that the family would be responsible. The site was only $100, so any additional cost would be hundreds of dollars less than the price at other cemeteries.

"In conservation cemeteries like Carolina Memorial Sanctuary, they mound the dirt on the grave, so it settles," I added. "Then the site never caves in at all."

"Well, that wouldn't work for us, because we have to mow," said Ray, who had a valid point. Lawn maintenance drives many burial practices, including concrete vaults, which keep the ground level. In conservation cemeteries, there is no grass to mow, because the habitat is natural. But in hybrid cemeteries, the lawn is a part of the aesthetic, for better or worse.

"In the years since Sheriff's burial, has the ground settled? And what have you done to maintain it?" asked John.

Ray proceeded to detail the inconveniences of mowing the cemetery, the difficulty of getting help, and the lack of a water source. When pressed, he admitted that additional care of the grave had been minimal, in part because the student had been buried in a shroud.

"I'll be buried in a shroud," I said, eager to agree. I had my organic shroud picked out and ready for order!

"Sheriff's case is entirely different from yours," Ray said.

No one said a word. After a long pause, I responded, "I don't think it's that different. We both held beliefs about life, death, and the earth based on our religious convictions."

"That's not how I see it," he said.

We'd come to an impasse, but at least the conversation had tackled some critical questions, including who would dig the grave. I'd suggested we consult with Cassie, the alum from the conservation cemetery, who'd started a business to help people plan their own funerals.

At this point, Steve suggested we let the committee convene for their deliberations.

"Why don't we go out to the cemetery to talk about it?" said Bill. "We could keep our distance."

To my surprise, Ray chimed in. "I could go right now," he said.

"I gotta get my shoes on!" said John from his black box.

As soon as I closed my laptop, Steve called me at home.

"Ray's not ever going to approve a green burial," I said.

"There are cemeteries that are more inclusive and accepting of different spiritual practices," he said. "It's just sad that we aren't one of them yet."

Later that day, Bill emailed me: "We indeed had a very productive meeting, and I anticipate you'll be pleased with the result. I have a question. Is there a standard practice in casketless burials regarding depth? Surely, it must be deep enough to keep critters from digging up the body, but I assume that too deep below the biologically active layer would also be less desirable."

I wrote back that green burial—shrouded or casketed—usually involved digging a hole about three to four feet, the ideal depth for decomposition.

The next day, Steve left a voice mail: "The trustees made a decision," he said. "They actually voted *not* to grant a one-time exemption."

I sighed, disappointed and confused, until he continued: "But they voted to *change* the policy. The committee is going to allow natural burial for those who want to minimize their environmental impact in the Warren Wilson Cemetery. This is the right thing to do. It's right for our community."

Then he took a deep breath: "You are good to go, as it were! And when I say 'good to go,' I mean that figuratively, of course."

About a month later, Steve emailed me with news: Ray had taken a sudden and rather precipitous downturn due to cancer. He wasn't expected to make it to the end of the week. His daughters were keeping vigil with hospice in his home.

It was late summer of the pandemic. In the middle of the night, I'd bolt awake in bed to worry about teaching both in person and online at the college while parenting a high school student learning on her laptop at home. I became addicted to the headlines: Cases of COVID-19 in the United States were rising every day. The election loomed. It was hard to have faith in much of anything besides multiple cups of coffee in the morning and my one beer at night.

"Oh, shit, I'm so sorry to hear this news," I wrote back to Steve. After expressing concern for Ray's children, I mentioned that Steve might want to find that yellow poster board—the unofficial database of the Warren Wilson Cemetery—somewhere in Ray's house.

Steve had one phrase for me: "O ye of little faith," he wrote. It turned out Ray had asked the college archivist to digitize the names and plots and preserve the yellow poster board in perpetuity.

"Apparently, he saw or maybe felt this day was coming," Steve said.

Ray died that week on July 12, 2020, but not before explaining to the cemetery trustees how to use the official tape measure and pins to mark the plots.

"Don't let anything happen to these," he'd said.

I'd learn later that as the trustees had walked around the cemetery after our Zoom meeting, Ray had been the one to propose changing the policy, rather than simply granting an exemption, which confounded us all. Soon after his death, his family set up a fund in Ray's name to maintain the cemetery, a legacy rooted in the land and our community.

In the beginning of this story, I asked, "What can I do?" But in the end, the bigger question became, What can we do together? As my mother always told me, we don't have to understand a mystery to experience one. But I was good to go—to purchase that plot and the shroud and to write the instructions my daughters would need someday.

Rather than simply filling out a legal form, I also decided to type up a plan like the one my father had given me. After the year, I saw the importance of revisiting these instructions for the people and the places in my life. Trapped in our house in a pandemic with mounting deaths each day, my children didn't want to talk about my funeral. But I read the words to them, just as my father had done for me after my mother died. His words found their place in my body, and mine would find a home in theirs someday. This might be my small part in a resurrection story.

My Death Plan

Dear Maya and Annie Sky,

Before I die, I want to share a road map with you. My father ensured I could grieve my mom's death by honoring her life, and he gave me a plan for the end of his life too. Like my parents, I hope for a natural burial without embalming or a vault in the ground, so my death doesn't harm the earth that has given us life.

First, my preference is for my body to stay at home if possible after I die. You can call Caroline Yongue at the Center for End of Life Transitions. (There is a list of phone numbers at the end of these instructions.) If she is not available, contact Trish Rux or Cassie Barrett. You won't have to take care of my body by yourself, and they can give guidance with logistics like finding a death midwife. (Then you don't have to go to the Motel 6 if you don't want to.)

I would like my body to stay at home for one to two nights with the burial on the second or third day. If that's not possible, call Stanley Combs at Asheville Alternatives to pick up and refrigerate my body. Tell him you only need transport and refrigeration, not any other services.

For the memorial, ask Brian Cole to do the homily and have the service at All Soul's Episcopal Church (or wherever I'm worshipping at the time). Follow the same Celebration of Life service as my parents, with the hymns from their funerals and liturgy from the Episcopal *Book of Common Prayer*. After the service, I would like Steve Thompson or another friend with a truck to transport my body to the Warren Wilson Cemetery. You can use the linen shroud from Mourning Dove Studio. (If you want a casket, Steve would build one, or you can buy a cardboard one for $300 from Mourning Dove. My dad thought a refrigerator box would work fine.)

I have purchased my plot at the Warren Wilson Cemetery, and the trustees should let you pay to have the grave dug, or Steve and his friends will dig it. (On an impulse, I bought plots for you both as well, but hopefully, I won't be around to see their use.) Invite Rev. Steve Runholt or the current pastor of the Warren Wilson Presbyterian Church to host and welcome people. Cassie and Aditi have a business helping people plan their own funerals, and they could also find a grave digger for you. At the gravesite, ask Brian to offer prayers, and then Liz Teague could sing her original tunes like "You Carry My Love," as well as gospel tunes like "I'll Fly Away" as my body is placed into the grave.

Please have shovels by the graveside, so family and friends can close the grave. Make sure the children dig up some dirt and throw it into the hole, even with their hands. That's what they will remember. Depending on the season, you can place spring flowers or fall leaves on top of the grave. During a pandemic, it's hard to imagine a potluck after the burial, but call my circle of female friends: They will organize a gathering with food, as my Mom's friends did for us after her death.

If, for some reason, a burial at Warren Wilson Cemetery isn't feasible, purchase a plot from Carolina Memorial Sanctuary and follow this same plan at the conservation cemetery. If I die far away from home, cremate my body; if aquamation is available, that's preferable. You can then have the memorial service and bury my cremated remains at the Warren Wilson Cemetery when it's convenient. Purchase a flat native stone and engrave it with the date of birth and death: Mallory DeVane McDuff 01-19-1966–XX-XX-XXXX. (Obviously, you'd fill in the dates!)

Every person listed in this plan is someone I encountered during this journey; they have given their lives to help people live with death. I am grateful they might be able to help you too. Then you can do the same for someone else one day.

This past summer, I attended the funeral of my mom's only brother, Cecil, a seventy-eight-year-old Episcopal priest who'd directed the church camp started by my grandparents in Mississippi. Because of the pandemic, I watched the service via Facebook Live as a small group of relatives sat distanced on wooden benches by the lake. I have a photo on our bookshelf of my grandfather, also a priest, paddling a canoe on this same lake with my mom as a child in the boat. Through generations that come before us, we are shaped by the land—how we live and how we die.

Honoring my parents after their deaths—and living into their love for the earth—has been one of the defining, unexpected paths of my life. I experience my parents' love in my body nearly every day, and that feels like God to me, even when I'm scared about what will happen next. You too will carry my love everywhere you go. And if you can't follow this plan, remember what Annie Sky once said: "If Mom's dead, she might never know."

But you'll always know I love you.

Yours forever,
Mommy

GRATITUDE

To the Tsalagi/Cherokee who lived and buried in this valley within their traditional homeland. To the forests, farms, mountains, and rivers of this region and all life sustained by this place. To each and every person who allowed me to spend time in funeral homes, attend the burial of a family member, enter the sacred space of a home vigil, or ask questions in coffee shops about death and dying. To the funeral directors, end-of-life doulas, cemetery managers, researchers, volunteers, and friends in this story, especially those who pointed out my assumptions and biases but continued to teach me. To all who are working to envision and create a just world beyond today.

To Caroline Yongue and the power of moving past our fears, especially when it's hard. To Steve Runholt and the trustees of the Warren Wilson Cemetery, especially Bill Sanderson and, of course, Ray Stock, who knew how to leave a legacy. To my students at Warren Wilson College—and to teaching and learning as an act of faith. To Roberta Zeff at the *New York Times*, who published the essay "Rest Me in a Pine Box and Let the Fiddle Play," from which the introduction to the book was adapted.

To my writing mentor, Jill Rothenberg, for feedback and friendship. To Liz Teague for lyrics and love. To my close friends and readers for their generosity and humor, especially Lyn O'Hare, Margaret Lewis, and Brian Cole. To literary agent Carol Mann and to editor Lil Copan and everyone at Broadleaf Books for letting me be a part of their vision. To Annie Sky and Maya for being exactly who they are. Lastly, to Ann and Larry McDuff, who showed me that love is the enduring connection between life, death, and earth.

GLOSSARY

Definitions related to green burial are adapted with permission from Lee Webster and Merilynne Rush, "Glossary of Green Burial Terms" (Green Burial Council, 2015), www.greenburialcouncil.org, updated 2019 by Lee Webster, Heidi Hannapel, Laura Starkey, and Caroline Yongue in a collaboration between the Green Burial Council and the Conservation Burial Alliance.

Alkaline hydrolysis—Also known as aquamation; water cremation; flameless cremation. A process of disposal of human and animal remains using water, lye, heat, and pressure to promote decomposition, leaving bone fragments and an effluent.

Biodegradable—Capable of breaking down into natural materials in the environment without causing harm; capable of being decomposed by bacteria or other living organisms.

Body donation—Donation of a body after death for purposes of education and research, often in the fields of medicine or science.

Body farm—An outdoor laboratory or facility used for the study and research of decomposition.

Burial containers—Caskets and shrouds capable of being decomposed or biodegraded by bacteria or other living organisms; often made of plant or animal fiber (wicker, sea grass, paper, linen, cotton, wool, willow, bamboo, etc.). Metals, glues, resins, plastics, and other synthetics that are nonbiodegradable are not recommended in green burial.

Burial ground—Also known as cemetery; preserve. An area set aside for the burial of human bodies or remains.

Burial plot—Also known as burial space. The space in which a body is buried.

Caskets—Containers for the dead, previously called coffins. The terminology appears to have evolved as a marketing tool to emphasize the precious cargo. "Casket" (from Middle English *casse* and Anglo-Norman French *cassette*) was originally used to denote a small ornamental box, case, or chest for carrying jewels, letters, or other valuable items. Conventional caskets are built of steel, copper, and other metals, fiberglass, and exotic woods. Many are dressed with symbolic or religious icons, jewels, engravings, fittings, or trimming (fabric lining).

Coffins—Six- or eight-sided containers for the dead used for burial or cremation. Eight-sided coffins, also called "toe pinchers," may be designed to conserve wood or to emphasize the shape of the human inside (wide shoulders tapering to small feet). Plain pine boxes tend to be thought of as coffins, though there is no limitation. The term is derived from the Greek word *kophinos*, meaning basket.

Conservation—The act of preserving, protecting, or restoring the natural environment, natural ecosystems, vegetation, and wildlife.

Conservation burial ground—A type of natural cemetery that is established in partnership with a conservation organization and includes a conservation management plan that upholds best practices and provides perpetual protection of the land according to a conservation easement or deed restriction.

Conservation easement—A voluntary legal agreement between a landowner and a land trust (or government agency) that permanently limits the uses of the land to protect its conservation values.

Conservation organizations—Nonprofits and governmental entities organized to acquire, monitor, and manage land, rivers, forests, and other natural resources to preserve and protect them through prudent management.

Conventional cemetery—Also known as lawn cemetery; modern cemetery. A cemetery that requires the use of a concrete or fiberglass grave liner or a vault and a hard-bottom casket. Prior to the establishment of modern cemeteries, most burial occurred in churchyards or on family land and was environmentally friendly. Modern cemetery requirements are dictated by "convention," rather than law.

Cremation—The process of reducing the body of the deceased to bone fragments and ashes by the use of high heat.

Death doula—Also known as end-of-life doula, death midwife. A nonmedical professional trained to care for a person who is dying, much like a birth doula's role during childbirth. Many people make the distinction of a home funeral guide or death midwife as a person also trained in caring for the body after death, such as during a home funeral.

Decomposition—The breakdown of the body by natural means (soil, water, heat, and microbes in balance). Natural decomposition, the goal of green burial, occurs when no chemicals or nonbiodegradable elements (steel, resins, fabrics, cement vaults) impede the process or attempt to preserve the body.

Disposition—Also known as final disposition. The placement of the remains of a body in their final resting place.

Dry ice—The solid form of carbon dioxide; may be used to cool and preserve a body temporarily during a home funeral (home vigil). Dry ice must be handled carefully to avoid skin burns and requires good ventilation due to off-gassing of carbon dioxide.

Embalming—The process of removing blood and fluids from the dead body and inserting preservatives, surfactants, solvents, and coloration to slow decomposition and improve looks for a period of up to two weeks. Organs are punctured and drained of fluid with the use of a sharp instrument called a trocar; waste is disposed of in a standard septic system or municipal wastewater treatment plant.

Embalming fluid—An array of chemicals, including benzene, methanol, ethyl alcohol, and ethylene glycol (antifreeze). Formaldehyde, which constitutes anywhere from 5 to 29 percent of the solution, is associated with increased risk of ALS (amyotrophic lateral sclerosis), leukemia, lymph hematopoietic malignancies, and brain cancer in embalmers.

Five Wishes—A program affiliated with the organization Aging with Dignity, providing an accessible advance directive document or a living will, called *Five Wishes*, that documents an individual's wishes for end-of-life care.

Grave liner—An outer burial container that covers only the top and side of the casket; usually made of plastic, fiberglass, or metal. *See* Outer burial container.

Green burial—Also known as natural burial. A way of caring for the dead with minimal environmental impact that aids in the conservation of natural resources, reduction of carbon emissions, protection of worker health, and restoration and/or preservation of habitat. Green burial necessitates the use of nontoxic and biodegradable materials, such as caskets, shrouds, and urns.

Green embalming fluid—A biodegradable, nontoxic, noncarcinogenic, and formaldehyde-free alternative to conventional embalming fluid. The process of embalming is the same regardless of which fluid is used.

Greenwashing—The act of deceptively marketing goods or services by hiding dubious aspects of their environmental profile. In the case of green burial, the full picture of environmentally sound practices is important. Using a casket of organic materials but made by using fossil fuels and child labor and transported 3,000 miles to its destination is not considered "green" (environmentally sound).

Home burial—The practice of full-body interment on residential land, usually in a rural setting. Local zoning and health department regulations apply, as do state-approved setbacks for known sources of water, buildings, and highways. Often these are considered family cemeteries and must be established and reported as such to government agencies and are usually restricted to blood relatives or extended family.

Home funeral—Also known as a DIY funeral. The process of family and friends, next of kin, or a designated agent retaining custody and control of the body for the time between death and disposition (burial or cremation); similar to a home vigil. A home funeral involves bathing and dressing the body and using dry ice, Techni Ice™, or other cooling mechanism as a preservative; it commonly lasts one to three days. A home funeral guide may provide education and support either prior to or during this time period.

Home vigil—The practice of family and friends sitting with the body continuously while the body is lying in honor in the home; also may simply refer to the time from death to disposition. A home vigil is similar to a home funeral; the terms may be used interchangeably.

Human composting—A process of accelerated decomposition that transforms human remains into soil. In the United States, human composting is legal in the State of Washington, Oregon, and Colorado.

Hybrid burial ground—A cemetery that allows vaults but also offers green burial.

Interment—Burial of the full body or the cremated remains of the deceased in a grave.

Islamic burial—Any burial practices common to the Islamic faith, which vary depending on the sect. These practices include collective bathing of the body, shrouding of the body, prayer (*salah*), unfettered burial of the shrouded body in the grave within twenty-four hours, and positioning of the head facing toward Mecca. Cremation is forbidden to Muslims.

Jewish burial—Any of a number of burial practices common to the Jewish faith, including burial within one day of death, wrapping the body in a white linen shroud made without knots, using a plain wooden coffin containing no metal, and providing for direct contact with the earth (achieved by drilling holes in the bottom of the coffin, using a bottomless vault, or having a green burial).

Mushroom suit—A biodegradable garment with the aim of digesting the body with the aid of mushroom spores and eliminating pesticides and heavy metals, among other contaminants.

Natural burial ground—Also known as green burial ground; green cemetery. A type of cemetery that allows full body interment in the ground, without embalming, using a biodegradable container, and without a grave liner or vault. Cremated remains and pet remains may be accepted in natural burial grounds.

Outer burial container—Either a burial vault or a grave liner that encases a casket or shrouded body. Both are used to support the soil around the casket from subsidence in most nongreen burial cemeteries to minimize cemetery maintenance by keeping the lawn flat for mowing. *See* Vault; Grave liner.

Shroud—Fabric or a sheet that is wrapped around the deceased for burial.

Techni Ice™—An effective, nontoxic, reusable dry-ice replacement used to cool the body of the deceased. It is purchased in plastic sheets, activated, and frozen. Unlike dry ice, it does not off-gas or cause vapors or condensation, and it can be reused indefinitely.

Vault—A container made of concrete, plastic, or metal that encloses a coffin or casket to help prevent a grave from sinking and provide some protection from the elements. The vault is installed into the grave. At the burial, the casket is placed inside the vault and sealed. Generally, outer burial containers are not required by state or local laws, but they are often required by conventional cemeteries to prevent the grave from collapse due to heavy maintenance equipment and ground settling. A vault may also be referred to as a burial vault, grave vault, or cemetery vault. *See* Outer burial container.

RESOURCES

The following books and websites provide a place for exploring next steps. See the notes for background research from each chapter.

Doughty, Caitlin. *From Here to Eternity: Traveling the World to Find the Good Death.* New York: Norton, 2018.

———. *Smoke Gets in Your Eyes and Other Lessons from the Crematory.* New York: Norton, 2014.

Fournier, Elizabeth. *The Green Burial Guidebook: Everything You Need to Plan an Affordable, Environmentally Friendly Burial.* Novato, CA: New World Library, 2018.

"Five Wishes for Individuals and Families." *Your Living Will and Advance Directive*, Five Wishes, 2021, https://tinyurl.com/tmu93m9p.

Green Burial Council, 2021. https://www.greenburialcouncil.org/.

Herring, Lucinda. *Reimagining Death: Stories and Practical Wisdom for Home Funerals and Green Burials.* Berkeley, CA: North Atlantic, 2019.

Johnson, Ayana Elizabeth, and Katharine K. Wilkinson, eds. *All We Can Save: Truth, Courage, and Solutions for the Climate Crisis.* New York: One World, 2020.

Long, Thomas G., and Thomas Lynch. *The Good Funeral: Death, Grief, and the Community of Care*. Louisville: Westminster John Knox, 2013.

Miller, BJ, and Shoshana Berger. *A Beginner's Guide to the End: Practical Advice for Living Life and Facing Death*. New York: Farrar, Straus & Giroux, 2019.

Recompose, 2021. https://recompose.life/.

Tisdale, Sallie. *Advice for Future Corpses (and Those Who Love Them): A Practical Perspective on Death and Dying*. New York: Gallery, 2018.

Western Carolina University. "Forensic Anthropology Body Donation Information." Accessed December 2, 2020, https://tinyurl.com/kerks95h.

NOTES

Chapter 1: Matters of Life, Death, and Earth

4 ***"And everything to do with the fact that we're all going to die"***: Ross Gay, "Tending Joy and Practicing Delight," interview by Krista Tippett, *On Being*, July 25, 2019, https://tinyurl.com/46fp5bky.

7 ***"I also shared what I'd learned about conservation burial grounds like Prairie Creek Conservation Cemetery in Florida"***: Mallory McDuff, *Sacred Acts: How Churches Are Working to Protect Earth's Climate* (Gabriola Island, BC: New Society, 2012).

8 ***"the Green Burial Council defines cemeteries as green"***: "FAQs: Green Burial Defined," Green Burial Council, n.d., https://tinyurl.com/psvmbtf3, accessed August 1, 2020.

10 ***"Conventional burials in the United States generate enormous environmental costs"***: McDuff, *Sacred Acts*, 50.

11 ***"Nearly 54 percent of people in this country are considering a green burial"***: Sonya Vatomsky, "Thinking about Having a 'Green' Funeral? Here's What to Know," *New York Times*, March 22, 2018, https://tinyurl.com/96hh4k6p.

11 ***"Each year in the United States, conventional burials require"***: McDuff, *Sacred Acts*, 51.

11 ***"Today more than 50 percent of the US population chooses flame cremation"***: Ann Carrns, "What to Know When Choosing Cremation," *New York Times*, July 26, 2019, https://tinyurl.com/4d4f6br8.

12 ***"One of my former students, Kelsey Juliana"***: Youth v. Gov website, Our Children's Trust, 2020, https://www.youthvgov.org.

12 ***"patriarchal power structures that support our fossil fuel economy"***: Ayana Elizabeth Johnson and Katharine K. Wilkinson, eds., *All We Can Save: Truth, Courage, and Solutions for the Climate Crisis* (New York: One World, 2020).

12 ***"more than fifty million people died each year worldwide"***: Elizabeth Fournier, *The Green Burial Guidebook: Everything You Need to Plan an Affordable, Environmentally Friendly Burial* (Novato, CA: New World Library, 2018), 23.

12 ***"How we handle the bodies of those who die can influence the climate on various levels"***: McDuff, *Sacred Acts*, 49.

12 ***"Forty percent of households in the United States earn less than $50,000 a year"***: "Household Income Distribution in the United States in 2018," Statista, https://tinyurl.com/yzr4x8z7.

13 ***"writes Mark Harris, author of* Grave Matters"**: Mark Harris, *Grave Matters: A Journey through the Modern Funeral Industry to a Natural Way of Burial* (New York: Scribner, 2007), 186.

13 ***"Many people want to have an environmentally conscious end to their life"***: Elizabeth Fournier, *The Green Burial Guidebook*.

Chapter 2: The Documents Class

21 ***"The purpose of this course was straightforward"***: BJ Miller and Shoshana Berger, *A Beginner's Guide to the End: Practical Advice for Living Life and Facing Death* (New York: Farrar, Straus and Giroux, 2019); and Perri Peltz and Matthew O'Neil, directors, *Alternative Endings: Six New Ways to Die in America*, 2019, on HBO, https://tinyurl.com/cn86njan.

31 ***"As writer Terry Tempest Williams said"***: Terry Tempest Williams, *Erosion: Essays of Undoing* (New York: Farrar, Straus and Giroux/Sarah Crichton, 2019).

Chapter 3: Innovative Undertakers

36 ***"serve the living while caring for the dead"***: Thomas G. Long and Thomas Lynch, *The Good Funeral* (Louisville: Westminster John Knox, 2013).

36 ***"Black funeral directors were on the front lines of the civil rights movement"***: Suzanne Smith, *To Serve the Living: Funeral Directors and the African-American Way of Death* (Cambridge, MA: Belknap, 2010).

37 ***"if there was a clock in the home, it was often stopped at the time of death"***: Rebecca Elswick, "Funeralizing: Death Practices of Appalachia," *Appalachian Magazine*, April 27, 2019.

38 ***"In the United States, the average conventional funeral and burial costs $10,000"***: Laura Holson, "As Funeral Crowdfunding Grows, so Do the Risks," *New York Times*, June 5, 2018, https://tinyurl.com/cvbsnm6j.

39 ***"We would never walk into a dealership"***: Joshua Slocum as quoted in Sallie Tisdale, *Advice for Future Corpses (and Those Who Love Them): A Practical Perspective on Death and Dying* (New York: Gallery, 2018), 157–58.

40 ***"the megaliths of funeral care like Service Corporation International (SCI)"***: Tisdale, *Advice for Future Corpses*.

41 ***"A 2015 study showed that 17 percent of adults aged twenty to thirty-nine years old had used the internet to request or donate money for funerals"***: Holson, "As Funeral Crowdfunding Grows."

41 ***"GoFundMe reports that 13 percent of its campaigns in 2017 were memorials"***: Holson, "As Funeral Crowdfunding Grows."

41 ***"Jessica Mitford's* The American Way of Death*"***: Jessica Mitford, *The American Way of Death* (New York: Simon and Schuster, 1963).

42 ***"A 2017 National Public Radio investigation of the funeral home industry"***: Robert Benincasa, "You Could Pay Thousands Less for a Funeral by Crossing the Street," *All Things Considered*, NPR, February 7, 2017, https://tinyurl.com/4vm2zm9v.

43 ***"Until the mid-1800s, families in this country cared for their dead at home before burial in a nearby cemetery, farm, or the backyard"***: Lucinda Herring, *Reimagining Death: Stories and Practical Wisdom for Home Funerals and Green Burials* (Berkeley, CA: North Atlantic, 2019); and Mark Harris, *Grave Matters: A Journey through the Modern Funeral Industry to a Natural Way of Burial* (New York: Scribner, 2007).

43 ***"A man named Thomas Holmes developed embalming techniques"***: Caitlin Doughty, *Smoke Gets in Your Eyes and Other Lessons from the Crematory* (New York: Norton, 2014).

43 ***"Embalming is not widely practiced worldwide"***: Elizabeth Fournier, *The Green Burial Guidebook: Everything You Need to Plan an Affordable, Environmentally Friendly Burial* (Novato, CA: New World Library, 2018).

43 ***"In a period of only 150 years, the practice of a family caring for the body at home has been replaced by the funeral home"***: Herring, *Reimagining Death.*

47 ***"Modern-day embalming involves draining the blood and replacing it with a formaldehyde-based preservative"***: Harris, *Grave Matters.*

47 ***"The funeral director pushes a piece of metal called a trocar into the stomach"***: Doughty, *Smoke Gets in Your Eyes.*

48 ***"She describes the purpose of the industry to 'protect, sanitize, and beautify the body'"***: Caitlin Doughty, "The Corpses That Changed My Life," TED Talk, YouTube, November 3, 2016, https://tinyurl.com/ebp33jws.

48 ***"She shares a story of preparing a baby's body"***: Doughty, *Smoke Gets in Your Eyes.*

48 ***"But morticians may use a needle injector"***: Doughty, *Smoke Gets in Your Eyes.*

52 ***"In the United States, casket shipments from manufacturers had an estimated value of $456 million in 2020"***: James Hagerty, "Funeral Industry Seeks Ways to Stay Relevant," *Wall Street Journal*, November 3, 2016, https://tinyurl.com/3be294zy.

Chapter 4: Dying and Its Aftermath

59 ***"In most states, these methods of cooling the body"***: Lucinda Herring, *Reimagining Death: Stories and Practical Wisdom for Home Funerals and Green Burials* (Berkeley, CA: North Atlantic, 2019).

60 ***"I spread resources on my coffee table"***: "How to Arrange a Home Funeral," Funeral Consumers Alliance, https://tinyurl.com/5cr4bxtr; Lee Webster, *Planning Guide and Workbook for Home*

Funeral Families (CreateSpace, 2015); and Herring, *Reimagining Death*. Also see Joshua Slocum and Lisa Carlson, *Final Rights: Reclaiming the American Way of Death* (Hinesburg, VT: Upper Access, 2011); and Stephen Jenkinson, *Die Wise: A Manifesto for Sanity and Soul* (Berkeley, CA: North Atlantic, 2015).

62 ***"Through searching online, I learned that the state required filing a 'notification of death'"***: Jessica Gillespie, "North Carolina Home Funeral Laws," Nolo, 2019, https://tinyurl.com/r6kzk5m2.

64 ***"He later founded the International End of Life Doula Association"***: Chase Beech, "Death Doulas Guide the Way for Those Who Face the End of Life," *Religious News Service*, November 8, 2019, https://tinyurl.com/38tvxkj6.

65 ***"Trish's advice echoed some of what I was reading in* Advice for Future Corpses*"***: Sallie Tisdale, *Advice for Future Corpses (and Those Who Love Them): A Practical Perspective on Death and Dying* (New York: Gallery, 2018).

Chapter 5: When Death Protects the Land We Love

74 ***"72 percent of US cemeteries report increased demand for green burial"***: Sonya Vatomsky, "Thinking about Having a 'Green' Funeral? Here's What to Know," *New York Times*, March 22, 2018, https://tinyurl.com/96hh4k6p.

75 ***"natural burial is defined as burial of an unembalmed body in a biodegradable container without a vault"***: Christopher Coutts, Carlton Basmajian, Joseph Sehee, Sarah Kelty, and Patrice Williams, "Natural Burial as a Conservation Tool in the U.S.," *Landscape and Urban Planning* 178 (October 2018): 130–43, https://tinyurl.com/3nk864t7.

75 ***"But I'd read about a survey by the National Funeral Directors Association, in which only 48 percent of respondents were aware"***: Vatomsky, "Thinking about Having a 'Green' Funeral?"

76 ***"three categories of green cemeteries based on specific criteria"***: Elizabeth Fournier, *The Green Burial Guidebook: Everything You Need to Plan an Affordable, Environmentally Friendly Burial* (Novato, CA: New World Library, 2018), 75.

77 ***"The ground contains enough coffin wood to construct forty homes"***: Mark Harris, *Grave Matters: A Journey through the Modern Funeral Industry to a Natural Way of Burial* (New York: Scribner, 2007).

77 ***"The average price of a green burial is $5,000, as opposed to $10,000 for a conventional one"***: Tony Rehagen, "Green Burials Are Forcing the Funeral Industry to Rethink Death," *Bloomberg*, October 27, 2016, https://tinyurl.com/nhfknet8.

86 ***"I'd read a study in the journal* Urban Forestry and Urban Greening *that natural burial can contribute to a range of ecosystem services"***: Andy Clayden, Trish Green, Jenny Hockey, and Mark Powell, "Cutting the Lawn: Natural Burial and Its Contribution to the Delivery of Ecosystem Services in Urban Cemeteries," *Urban Forestry and Urban Greening* 33 (June 2018): 99–106, https://tinyurl.com/24edkt59.

86 ***"City planners even see conservation burial grounds as one tool for creating open spaces"***: Alexandra Harker, "Landscapes of the Dead: An Argument for Conservation Burial," *Berkeley Planning Journal* 25 (2012): 150–59, https://tinyurl.com/czv2ydhr.

87 ***"Of the 162 sites, 99 are hybrids, 54 are natural burial grounds, and 10 are conservation burial grounds"***: Coutts et al., "Natural Burial as a Conservation Tool in the U.S."

87 ***"the US practice of burying embalmed human remains in a lawnpark cemetery"***: Coutts, et al., "Natural Burial as a Conservation Tool in the U.S.," 130.

87 ***"A 2021 inventory revealed a total of 300 green burial cemeteries:"*** New Hampshire Funeral Resources, Education, and Advocacy, "Green Burial Cemeteries in the U.S. and Canada," May 30, 2021, https://www.nhfuneral.org/green-burial-cemeteries-in-the-us-and-canada.html

Chapter 6: Bury Me Close to Home

93 ***"green burial was also a way to talk about resurrection"***: Elizabeth Fournier, *The Green Burial Guidebook: Everything You Need to Plan an Affordable, Environmentally Friendly Burial* (Novato, CA: New World Library, 2018).

94 ***"In my readings, I'd learned that the word cemetery comes from the Greek"***: Sallie Tisdale, *Advice for Future Corpses (and Those Who Love Them): A Practical Perspective on Death and Dying* (New York: Gallery, 2018).

104 ***"The average cost of a burial plot"***: Kate Wright, "How Much a Burial Plot Costs + 4 Tips to Get a Fair Price," *Cake*, October 10, 2019, https://tinyurl.com/36m9sby4.

Chapter 7: The Container Store

112 ***"The coffin replaced the shroud because people believed"***: Mark Harris, *Grave Matters: A Journey through the Modern Funeral Industry to a Natural Way of Burial* (New York: Scribner, 2007).

113 ***"By the end of the 1800s, a simple pine box wasn't sufficient for those with resources"***: Harris, *Grave Matters*.

113 ***"From my research, I knew that no federal or state laws required purchasing a casket from a funeral director"***: Harris, *Grave Matters*.

114 ***"It was made of simple blue-gray linen by a local fashion-designer-turned-antique-dealer"***: Daniel Walton, "Shrouded House: Recycled Linen Burial Shrouds Bring a Delicate Topic down to Earth," *Asheville Made*, May 1, 2018, https://tinyurl.com/yk892md6.

115 ***"Shroud burial is an ancient tradition that was practiced by some early Romans and Hebrews"***: Harris, *Grave Matters*.

115 ***"To comply with Green Burial Council standards, a shroud must be made from biodegradable"***: Elizabeth Fournier, *The Green Burial Guidebook: Everything You Need to Plan an Affordable, Environmentally Friendly Burial* (Novato, CA: New World Library, 2018).

115 ***"Books like* The Green Burial Guidebook *include instructions for sewing a shroud"***: Fournier, *The Green Burial Guidebook*.

115 ***"As I'd learned at Carolina Memorial Sanctuary, a common practice is to tuck fresh flowers"***: Cassie Barrett, "Biodegradable Burial Containers for Green Burial: The What and Why and Shrouds," Carolina Memorial Sanctuary, July 22, 2019, https://tinyurl.com/2sv4e4f6.

117 ***"The YouTube video he'd attached was entitled 'How to Build a Coffin'"***: "How to Build a Coffin," Nature's Caskets, YouTube, March 2, 2015, https://tinyurl.com/nps5szk7.

119 ***"Biodegradable burial containers are more affordable, simpler, and better for the environment"***: Barrett, "Biodegradable Burial Containers for Green Burial."

119 ***"For my budget, it mattered that the average cost of a green burial container was $275 to $3,000"***: Barrett, "Biodegradable Burial Containers for Green Burial."

120 ***"I also found free online plans for building a simple pine box for burial"***: "DIY Coffin Plans," Piedmont Pine Coffins, accessed January 16, 2021, https://tinyurl.com/fbttexzc; "Build Your Own Casket," Northwoods Casket Company, March 10, 2021, https://tinyurl.com/2c3np87d.

Chapter 8: Ashes to Ashes, Dust to Dust

128 ***"Decades later, I learned about the town of Crestone, Colorado"***: Caitlin Doughty, *From Here to Eternity: Traveling the World to Find the Good Death* (New York: Norton, 2018).

128 ***"The cremation of one body produces 250 to 350 pounds of carbon dioxide, the equivalent of a thousand-mile car trip"***: Elizabeth Fournier, *The Green Burial Guidebook: Everything You Need to Plan an Affordable, Environmentally Friendly Burial* (Novato, CA: New World Library, 2018).

129 ***"Cremations in the United States produce 250,000 tons of carbon dioxide emissions"***: Greta Moran, "What's Greener than Burial or Cremation? Human Composting," *Grist*, December 31, 2018, https://tinyurl.com/ntn6rfsz.

130 ***"It turns out the small daily choices we make can have a collective impact on the climate"***: Greg Dalton, "Inconspicuous Consumption: The Environmental Impact You Don't Know You Have," *Climate One*, Commonwealth Club, January 10, 2020, https://tinyurl.com/yk6hbdyv.

130 ***"Only 100 companies in the world are responsible for 71 percent"***: Tess Riley, "Just 100 Companies Responsible for 71% of

Global Emissions, Study Says," *Guardian*, July 10, 2017, https://tinyurl.com/st4fspwc.

130 ***"the US military is the institution"***: Ayana Elizabeth Johnson and Katharine K. Wilkinson, eds., *All We Can Save: Truth, Courage, and Solutions for the Climate Crisis* (New York: One World, 2020).

130 ***"three-quarters of people in the United States heard someone talk about the climate only once or twice a year"***: Katharine Hayhoe "Why We Need to Talk about Climate Change," *Climate One*, Commonwealth Club, January 22, 2019, https://tinyurl.com/u39h4xh3.

131 ***"The percentage of people choosing cremation in the United States"***: Ann Carrns, "What to Know When Choosing Cremation," *New York Times*, July 26, 2019, https://tinyurl.com/4d4f6br8.

131 ***"This is a remarkable shift in US cultural practices"***: Fournier, *The Green Burial Guidebook*.

131 ***"In many other countries, cremation remains a popular choice as well"***: Annabelle Garwita, "Cremation Nation: Our New Way to Go," *Wall Street Journal*, March 29, 2019, https://tinyurl.com/m9svdka.

132 ***"The Funeral Consumers Alliance says a reasonable rate for direct cremation without additional services is $800 to $1,200"***: Carrns, "What to Know When Choosing Cremation."

134 ***"Before watching videos of bodies burning, I'd read accounts of cremation from funeral director Caitlin Doughty"***: Caitlin Doughty, *Smoke Gets in Your Eyes and Other Lessons from the Crematory* (New York: Norton, 2014).

134 ***"The story goes that the first coal-fired crematory in the United States was in Pennsylvania"***: Doughty, *From Here to Eternity*.

134 ***"The first link that popped up included bright-red text in large capital letters announcing"***: BILLSTMAXX, "The Process of a Cremation and a Crematorium. WARNING!!! GRAPHIC," YouTube, November 30, 2011, https://tinyurl.com/pdsjvxun.

134 ***"Today, not being forced to see corpses'"***: Doughty, *From Here to Eternity*.

135 ***"The process emits mercury into the air"***: Fournier, *The Green Burial Guidebook*.

136 ***"I'm still obsessed with that book called* Advice for Future Corpses*"***: Sallie Tisdale, *Advice for Future Corpses (and Those Who*

Love Them): A Practical Perspective on Death and Dying (New York: Gallery, 2018).

137 ***"It's legal in twenty states, including North Carolina"***: For a listing of states, see Valerie Keene, "Alkaline Hydrolysis Laws in Your State," Nolo, 2020, https://tinyurl.com/em6t44zb.

137 ***"I'd learned about the growing popularity of alkaline hydrolysis, or aquamation—especially since its total carbon footprint is one-tenth that of flame cremation"***: Jonah Bromwich, "An Alternative to Burial and Cremation Gains Popularity," *New York Times*, October 19, 2017, https://tinyurl.com/67smj4bj.

137 ***"I'd also read that alkaline hydrolysis produced 58 total kg of carbon dioxide"***: Fournier, *The Green Burial Guidebook*.

138 ***"The costs are about $150 to $500 more than for flame cremation"***: Fournier, *The Green Burial Guidebook*.

138 ***"In some states, casket makers had lobbied against legalization"***: Emily Atkin, "The Fight for the Right to Be Cremated by Water," *New Republic*, June 14, 2018, https://tinyurl.com/yvvfdu3e.

138 ***"There are hidden costs as well"***: "Reconsider the Ecological Impact," Funeral Consumers Alliance of the Virginia Blue Ridge newsletter, Fall 2019; and Emily Atkin, "The Fight for the Right."

Chapter 9: Giving Back

151 ***"In 2017,* Esquire *magazine featured the story of a local woman, Kate Oberlin"***: Libby Copeland, "Kate's Still Here," *Esquire*, November 15, 2017, https://tinyurl.com/mxajj9ps.

152 ***"A single acre of soil contains 2,400 pounds"***: Caitlin Doughty, *From Here to Eternity: Traveling the World to Find the Good Death* (New York: Norton, 2018).

155 ***"When you add moisture and oxygen, temperatures inside a compost pile"***: Doughty, *From Here to Eternity*.

155 ***"The cost per person at the Seattle facility is $5,500"***: Brendan Kiley, "Recompose, the Human-Composting Alternative to Burial and Cremation, Finds a Home in Seattle's SoDo Area," *Seattle Times*, November 19, 2019, https://tinyurl.com/med3azvf; and Recompose website, 2021, https://recompose.life.

155 ***"During this monthlong process, nutrients in bodies after death support life"***: "Olson Kundig Reveals Recompose Facility in Seattle for Composting Human Bodies," Designboom, November 23, 2019, https://tinyurl.com/j5x28vma.

159 ***"In Mary Roach's book* Stiff*"***: Mary Roach, *Stiff: The Curious Life of Human Cadavers* (New York: Norton, 2003), 70.

163 ***"We are a loose collection glued briefly into a provisional thing called self, and all such things are bound to dissolution"***: Sallie Tisdale, *Advice for Future Corpses (and Those Who Love Them): A Practical Perspective on Death and Dying* (New York: Gallery, 2018), 205.

167 ***"The stages include distention, rupture"***: Doughty, *From Here to Eternity*, 165.

167 ***"During this research, I also read about similar options, including a technique called promession"***: Elizabeth Fournier, *The Green Burial Guidebook: Everything You Need to Plan an Affordable, Environmentally Friendly Burial* (Novato, CA: New World Library, 2018).

Chapter 10: Write It Down and Talk about It

170 ***"In this way, I learned about a nonprofit called Narrow Ridge Center"***: Justin McDuffie and Maggie Gregg, "No Casket, No Cost: Tennesseans Go Back to 'Natural' Burial," *Tennessean*, February 19, 2020, https://tinyurl.com/z92vku6y.

177 ***"More people were documenting their end-of-life wishes, some using online services"***: Jennifer Miller, "Boom Time for Death Planning," *New York Times*, July 16, 2020, https://tinyurl.com/s3c2hbz4.

177 ***"At Larkspur Conservation, the staff gave each person a color-coded shovel"***: Alejandra Molina, "Months in COVID-19, Funeral Directors and Clergy Continue to Innovate Death Care," *Religion News Service*, July 10, 2020, https://tinyurl.com/52ywy6h6.

179 ***"Disasters often unfold like revolutions"***: Rebecca Solnit, "Who Will Win the Fight for a Post-coronavirus America?," *New York Times*, March 29, 2020, https://tinyurl.com/5n358xey.

184 ***"the bigger question became, What can we do together?"***: Ayana Elizabeth Johnson and Katharine K. Wilkinson, eds., *All We Can Save: Truth, Courage, and Solutions for the Climate Crisis* (New York: One World, 2020).